ENDLESS NOVELTIES OF EXTRAORDINARY INTEREST

DOUG MACDOUGALL

Endless Novelties of Extraordinary Interest

The Voyage of

H.M.S. Challenger *and*

the Birth of

Modern Oceanography

Yale UNIVERSITY PRESS

NEW HAVEN AND LONDON

Published with assistance from the foundation established in memory of Calvin Chapin of the Class of 1788, Yale College.

Yale University Press books may be purchased in quantity for educational, business, or promotional use. For information, please e-mail sales.press@yale.edu (U.S. office) or sales@yaleup.co.uk (U.K. office).

Set in Janson type by Westchester Publishing Services.
Printed in the United States of America.

Library of Congress Control Number: 2019931293
ISBN 978-0-300-23205-9 (hardcover : alk. paper)

A catalogue record for this book is available from the British Library.

This paper meets the requirements of ANSI/NISO Z39.48-1992 (Permanence of Paper).

10 9 8 7 6 5 4 3 2 1

For curiosity, science, and the pursuit of truth

CONTENTS

Just over a century and a half ago, six scientists with an overwhelming curiosity about the natural world set off from Portsmouth, England, on an adventure that would occupy them for the next three and a half years. They did not depart on their own, however; they were embedded with more than 250 sailors and officers of the British Royal Navy on a ship called H.M.S *Challenger*. The story of their voyage is one of discovery and adventure, hardship and humor. The expedition they embarked on was, uniquely, a product of its place and time. In hindsight it is easy to take a historical enterprise like this for granted, as something that was bound to happen. But in its day it was revolutionary.

My first encounter with the *Challenger* expedition came about serendipitously. I was a new graduate student at the Scripps Institution of Oceanography in La Jolla, California, searching for a suitable Ph.D. research project (in those heady days we were expected to devise our own; there were few, if any, ready-made projects to take up). I was particularly interested in the sediments of the deep sea floor, and one day in the library, as I glanced at the pile of books I had pulled from the stacks, I noticed one that stood out. It was both larger and older than the others, and the title on the spine said simply, *Deep-Sea Deposits*. Inside I found a more formal and much longer title: *Report on Deep-Sea Deposits Based on the Specimens Collected During the Voyage of H.M.S. Challenger During the Years 1872 to 1876*. The authors were John Murray and the Rev. A. F. Renard, and the publication date was 1891. The volume was "Published by Order of Her Majesty's Government," the title page said.

In an editorial note at the beginning of the book, John Murray, the lead author, explained that he hoped the information it contained would be useful because it was "the first attempt to deal systematically with Deep-Sea deposits, and the Geology of the sea bed." The volume contained hundreds of pages of detailed descriptions of sea-bottom sediments from almost every corner of the world, and the text was accompanied by exquisite drawings of things recovered from the ocean floor: minerals, sharks' teeth, manganese concretions, shells, and skeletons of creatures such as foraminifera and diatoms. (Photography was relatively new and cumbersome at the time, so although some specimens were photographed, many more were depicted in detailed, precise sketches.) I was entranced.

The book I was looking at, it turned out, was just one of fifty similarly massive volumes of the Challenger Report. (You read that correctly, there really are fifty.) They comprise the official record of the groundbreaking expedition, detailing its scientific findings and bringing together descriptions of all the specimens recovered. Many scientists, not just the six who had participated in the expedition, were involved in the work that went into producing the report, and the final volume did not appear until 1895, nearly twenty years after the voyage ended. Over the years since my first encounter with these volumes I have periodically returned to them, and to other material related to the expedition. Why such interest in a voyage that took place so long ago? Aside from the science itself—which was certainly important, but to a large degree has been overtaken by modern research—I am fascinated by the question of why the expedition has attained such an iconic place in the annals of ocean exploration and science. Why have so many later vessels of discovery—such as the lunar module piloted by Harrison Schmidt on the *Apollo* 17 mission and a space shuttle—been named after H.M.S. *Challenger?* And what was it that compelled those six *Challenger* scientists to leave their offices, their laboratories, and the comforts of home to embark on such a long and at times difficult sea voyage?

The *Challenger* expedition was first and foremost about science. But it was funded by the British government through the navy, which turned over one of its vessels to this small group of scientists to sail around the world for several years making measurements and collecting samples of everything from seafloor mud to plants and animals from the islands they visited. Furthermore, the navy provided funds for their salaries as well as for the microscopes, sample jars, dredges, and other pieces of scientific equipment they would need on the voyage. Why? Part of the reason, I believe, had to do with the expedition's timing, as will be explored in this book. The Victorian era was characterized by curiosity about the natural world—among scientists certainly, but also among the general public. Amateur naturalists abounded: birdwatchers, fossil hunters, butterfly collectors. Some of these nature enthusiasts donated their personal collections to newly established museums of natural history or even set up their own museums. An expedition to explore the ocean was bound to make discoveries and bring back an abundance of new marine creatures for display. It was almost guaranteed to be a popular venture. That perception was borne out quickly: as the voyage progressed, dispatches from *Challenger* appeared in newspapers and periodicals worldwide and were eagerly awaited, especially in Britain. When *Challenger* finally returned home from her epic journey and moored at the navy docks near the mouth of the Thames, so many people wanted to visit the now famous ship that special trains had to be laid on to take passengers from London to the dockyards.

Beyond the intrinsic interest in exploration and discovery, compelling practical considerations came into play when the expedition was first proposed. Britain depended on sea power to guard its global trading routes and protect the far-flung colonies of the British Empire, which were the source of much of the country's wealth. New information about the vast tracts of ocean patrolled by the Royal Navy would be inherently valuable. In addition, much of the burgeoning business of laying deep sea telegraph cables was centered in Britain, and *Challenger*'s characterization of the largely unknown deep sea floor would be a boon to that industry. Britain was powerful, supremely confident, and the richest country in the world; the government could afford to mount such an expedition. Not only would it enhance Britain's reputation as a seagoing nation and supporter of scientific inquiry; it would also have strategic value.

The *Challenger* expedition was not the first voyage of ocean exploration with a scientific component. British naval voyages, and those of other countries too, often included a naturalist. And for decades, serving officers in the British navy had been tasked with taking measurements of the earth's magnetic field, recording meteorological data, observing the flora, fauna, and geology of remote lands, and surveying coastlines. Some of these officers had been recognized for their scientific work and became prominent members of the Royal Society, an influential body of scientists and intellectuals. Before the *Challenger* expedition, however, most of the navy's peacetime voyages of exploration had been geographically limited and focused on practical or commercial matters, such as surveys of shipping routes, acquisition of new territories, or the search for a northwest passage to the Orient. The scientific observations also concentrated on the practical: magnetism for navigation, flora and fauna as potential food sources, geology for possible mining ventures. But crucially, this meant that the navy and its officers were well prepared for an expedition like that of *Challenger*, concerned solely with science. I have highlighted here only Britain, but a few other countries had similar legacies of scientific expertise in their navies, and by the 1870s there were indications that the United States and Germany in particular were making more ambitious plans for ocean exploration. This, too, was potent motivation for the British government to take on a project like the *Challenger* expedition. The country could not risk being left behind.

The ambitious plans drawn up for the voyage marked it as the world's first global oceanographic expedition driven purely by scientific curiosity. The proposed itinerary would take the ship to most parts of the globe, even the Antarctic, and it explicitly included visits to remote oceanic islands and provisions to collect and study samples of rare terrestrial plants and animals. In some ways the expedition was the *Apollo* project of its day: government-funded "big science" that captured the public imagination. Before 1872, when *Challenger* set out on her historic journey, the word *oceanography* did not appear in any dictionary. But by the time the ship returned it was obvious that multidisciplinary study of the oceans was a field of science in its own right. John Buchanan, one of the scientists on board, later claimed that the discipline of oceanography was born on February 15, 1873. That was the day the first official observing station of the expedition was occupied, in the Atlantic Ocean west of the island of Tenerife.

Oceans cover more than two-thirds of the earth's surface. They have a major impact on the planet's climate and weather; the global effects of El Niño, for example, have attracted much media attention in recent years. Photosynthesis by organisms that live in the sea produces, by most estimates, more than two-thirds of the oxygen we breathe. Marine organisms that make their shells from calcium carbonate help regulate the amount of carbon dioxide in the atmosphere. Oceans give us food and other resources and are an early-warning system for global warming. Yet in spite of their importance to human life, they are so vast that many parts of them remain little known even today. That was doubly true at the time of the *Challenger* expedition; as Charles Wyville Thomson, the expedition's scientific leader, wrote in his book *Depths of the Sea*, the deep ocean was the only remaining place on earth where naturalists could expect to find "endless novelties of extraordinary interest."

In this book I examine Britain's early attempt to explore and understand the oceans, not as a chronological account of the expedition—several other books do that well and I list them in the bibliography—but rather through a series of topics that I think shed light on some of the important discoveries of the expedition, as well as on what drove the civilian scientists on board, those quintessentially curious Victorian intellectuals, in their pursuit of science at sea. Their exploits deserve to be more widely known, and I have tried to tell the story of the voyage through their experiences. I hope you enjoy the journey.

CAST OF CHARACTERS

To help readers follow the narrative, I list here, in alphabetical order by last name, the people on the *Challenger* expedition who make more than a cameo appearance in this book or whose writings are quoted at some length. For clarity I distinguish between civilian and naval personnel.

John Buchanan: chemist (civilian)
George Campbell: sublieutenant (navy)
Joseph Matkin: steward's assistant (navy)
Henry Moseley: naturalist (civilian)
John Murray: naturalist (civilian)
George Nares: ship's captain for first part of the expedition (navy)
William Spry: engineer (navy)
Herbert Swire: navigating sublieutenant (navy)
Charles Wyville Thomson: naturalist and scientific director (civilian)
Frank Thomson: ship's captain for last part of the expedition (navy)
Thomas Tizard: navigating lieutenant and senior surveying officer (navy); later helped write the Challenger Report
John Wild: secretary to Thomson and expedition artist (civilian)
Rudolf von Willemoes-Suhm: naturalist (civilian)

A few other men did not participate in the expedition but appear in the book in connection with it.

William Carpenter: zoologist who worked with Wyville Thomson to secure government support for the *Challenger* expedition
Ernst Haeckel: German naturalist who helped popularize Charles Darwin's work in Europe and reported on parts of the *Challenger* collection of marine organisms

Thomas Huxley: prominent naturalist and intellectual who championed Darwin's evolutionary ideas; sometimes known as "Darwin's bulldog"

Alphonse Renard: Belgian geologist who helped John Murray with his studies of rock and sediment samples at the conclusion of the expedition

George Richards: hydrographer of the navy and fellow of the Royal Society who was instrumental in gaining Admiralty approval for the expedition

ACKNOWLEDGMENTS

Thanks go to my agent Rick Balkin for his persistence in shepherding this project over several hurdles, and for his incisive comments on the manuscript. Also to Anthony Arnove, who took over some of the reins and dealt with a number of administrative matters. At Yale University Press my heartfelt thanks especially to Joe Calamia and Susan Laity, Joe for his initial enthusiasm for this project, his encouragement along the way, and his perceptive editorial comments, which significantly improved the manuscript; and Susan for her meticulous attention to detail, which transformed the manuscript into a better book. Two anonymous readers provided much useful feedback on an early draft. I also want to single out the helpful staff at the Centre for Research Collections at the University of Edinburgh, who patiently dealt with my requests as I sifted through multiple boxes of original *Challenger* papers. And as always, thanks to Sheila, my first and chief reader.

Maps

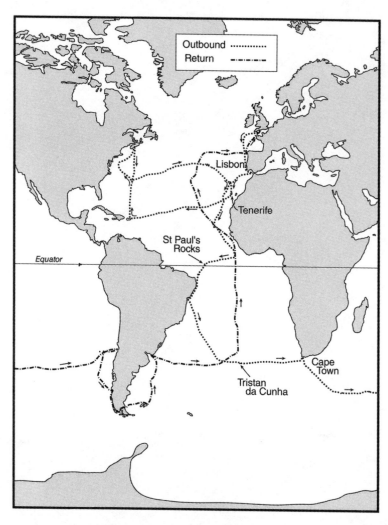

Challenger's track in the Atlantic Ocean,
December 1872–December 1873 and January–May 1876.

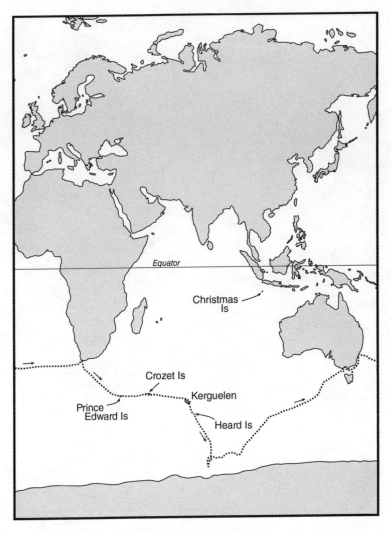

Challenger's track in the Southern Ocean, December 1873–March 1874.

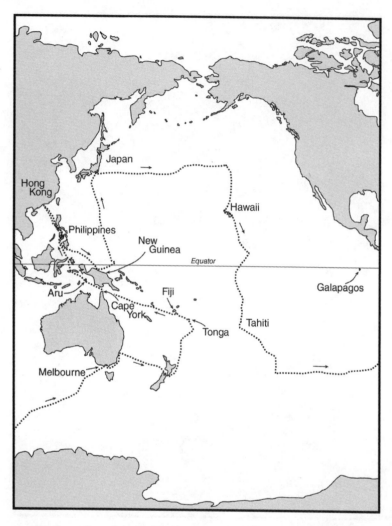

Challenger's track in the Pacific Ocean, March 1874–January 1876.

ENDLESS NOVELTIES OF EXTRAORDINARY INTEREST

one COSMIC DUST

Shortly before 8:45 on the evening of December 21, 1872, Phileas Fogg, the hero of Jules Verne's novel *Around the World in Eighty Days*, went to his London club to claim a twenty-thousand-pound reward for circumnavigating the world. Months earlier he had made a bet that he could complete the journey in eighty days. None of his friends believed he could do it; they were quite happy to take on a wager they were sure they could win. In fact, at the end of his journey Fogg himself thought he had lost. He had kept careful track of his days, and according to his calculations he was one day late. He was despondent; he had lost the bet by the narrowest of margins. But since this was a novel with a happy ending, it turned out he had missed something: he had forgotten to factor in the international dateline, which he had crossed during his travels, gaining one calendar day. In London it was not the 22nd of December as he believed, but the 21st, and he had indeed completed the journey in eighty days. When he realized this Fogg rushed to his club at the last minute, a happy—and soon to be very rich—man. In today's terms his winnings would equal over two million pounds, or roughly three million U.S. dollars.

On the same day, December 21, 1872, as the fictitious journey depicted in Verne's novel came to an end, a real and more momentous voyage around the world was just beginning: the Royal Navy ship H.M.S. *Challenger* was setting out from Portsmouth—not that far from Phileas Fogg's London—on what was to be the world's first major global oceanographic expedition. The crew and scientists aboard *Challenger* would circumnavigate the globe, as Fogg had done, but their journey would last much longer than 80 days. *Challenger* did not return to England for more than three years—1,250 days later, to be exact, on Queen Victoria's birthday, May 24, 1876. By then the ship had crisscrossed the Atlantic, rounded both the Cape of Good Hope and Cape Horn, sailed south toward Antarctica until her way was blocked by ice, and visited numerous remote islands in the Atlantic, Pacific, and Southern Oceans. And it too had crossed the dateline. (Joseph Matkin, assistant to the ship's steward, wrote to his mother that the week of the *Challenger*'s dateline crossing had eight days but the Admiralty would only pay wages for seven. However, he said, the extra day was declared a Sunday and the sailors were given a holiday.) When *Challenger* finally arrived back in England no one on board could claim a monetary reward like Phileas Fogg's. But for many their experiences on the long voyage were compensation enough.

The *Challenger* expedition was the brainchild of two men who had a common interest in the biology of the oceans and had worked together in the past: Charles Wyville Thomson, professor of natural history at Edinburgh University, and another prominent scientist, William Carpenter, who was based in London. Both were members of the Royal Society, at the time an immensely influential organization, and they submitted their plan for ocean exploration to the Admiralty, which oversaw the navy, under the official auspices of the society. They had a powerful ally in the hydrographer of the navy, a fellow member of the Royal Society, and with surprisingly little hesitation, given the magnitude of the project, the Admiralty agreed to the proposal. In monetary terms the project would turn out to be by far the most expensive single scientific endeavor ever attempted, and it remained one of the largest on record until the mid-twentieth century.

Once it had been agreed that the expedition would proceed, the Royal Society put together a Circumnavigation Committee: a group of prominent scientists whose charge was to set out *Challenger*'s course and outline the most important tasks she should undertake. It was agreed from the beginning that the voyage was to be dominated by science rather than the needs of the navy, but during the planning stages frequent communication, mostly in the form of written correspondence, passed between the Royal Society and the Admiralty. The Admiralty had questions: what were the "precise objects of research" for the expedition, and "in what particular portions of the ocean" would the investigations "be carried out with the greatest advantage to science?" The Circumnavigation Committee replied in overwhelming detail, setting out plans for a wide-ranging investigation that would include determining the physical, chemical, and biological characteristics of all the world's oceans, with special emphasis on the deep sea. The documents submitted to the Admiralty constituted an extensive wish list, the kind of proposal that most scientists today could only dream of.

The committee explained at length precisely why certain measurements were necessary and described how they should be carried out. The section of the correspondence that dealt with one small task, the collection of marine algae samples, is a good example. It laid out in excruciating detail the scientific procedure: "The more delicate kinds [of algae], after gentle washing, may be floated in a vessel of fresh water, upon thick and smooth writing or drawing paper; then gently lift out paper and plant together, allow some time to drip then place on the sea-weed clean linen or cotton cloth, and on it a sheet of absorbent paper, and submit to moderate pressure—many adhere to paper, but not to cloth; then change the cloth and absorbent paper till the specimens are dry." We can imagine a dazed Admiralty bureaucrat reading this and wondering what in the world it had to do with naval operations.

The navy was responsible for selecting a vessel for the expedition, and it chose *Challenger*. Launched in 1858, she was not originally built for science; rather she was a small warship fitted with cannons and designed primarily for duties such as coastal patrols or the support of larger vessels in the fleet. But for her scientific voyage she was refitted and repurposed: onboard laboratories were constructed, accommodation was provided for the civilian scientists, and most of her guns were removed, although two were retained so that ceremonial salutes could be made in foreign ports. (One of the scientists, Henry Mosely, thought this practice was an anachronism. After observing a back-and-forth salute at a Dutch colony in Indonesia he wrote: "It is to be hoped, that before long the intolerable nuisance of saluting will be done away with; it is most astonishing that civilized persons can be so much the slaves of habit, as to make a painful noise of this

H.M.S. *Challenger* under sail. (Sepia drawing by J. J. Wild, courtesy of the Centre for Research Collections, University of Edinburgh.)

kind when necessity does not require it; every one concerned dislikes the noise, and there is a great waste of material.") Fitted out for research rather than aggression, *Challenger* was to have a new role during the years of the expedition, serving science rather than overtly defending British interests. In the process she became one of the best-known ships of the British navy.

William Carpenter and Wyville Thomson were both biologists. Although only Thomson sailed with the expedition—Carpenter remained ashore—both men anticipated that the voyage would result in many new discoveries relating to the flora and fauna of the oceans. They were not disappointed. But in their years pursuing science at sea, the *Challenger* naturalists made discoveries in other areas of science, too. One was documented in the Challenger Report volume mentioned in my Preface: *Deep-Sea Deposits*. A color plate in this thick book shows several small, spherical, dark-colored objects that immediately caught my attention when I first saw them, especially because the annotation on the facing page identified them as "cosmic magnetic spherules."

In terms of numbers or weight the "cosmic spherules" were a minor part of the treasure trove of materials that *Challenger*'s scientists brought up from the seafloor, but they nevertheless raised important questions. The first was what cosmic material was doing at the bottom of the ocean. It might seem an improbable place to find objects from outer space; could the expedition's scientists have been wrong? In fact, though, they were correct in their identification. Over the many decades since the expedition it has gradually become clear that parts of the deep sea floor are repositories of considerable quantities of extraterrestrial material. By targeting these areas, scientists have collected thousands of spherules like those found by *Challenger*. These fascinating objects have been probed and analyzed using every imaginable kind of scientific instrument, and there is no question that they are from space.

When I first read about the *Challenger* discovery I already knew a little about extraterrestrial particles on earth. One Christmas when I was quite young an aunt gave me a magazine subscription as a gift. The magazine was called *Boys' Life*, and it was—and is still—the official publication of the Boy Scouts of America. In those days it was full of articles about nature and the outdoors, and I awaited its arrival with great anticipation. Although most of the *Boys' Life* stories have long since disappeared from my memory, the details of one remain. It was a story about cosmic dust.

Cosmic dust! The words conjure up all sorts of images, especially in the imagination of a young boy. Our planet, the article's author explained, is continuously bombarded with extraterrestrial material, most of it in the form of dust-sized particles—pieces of exotic matter from some distant corner of outer space drifting down onto the earth. "Shooting stars," I learned, those brief flashes of light that streak across the night sky, are the visible traces of the tiny particles smashing through the atmosphere.

The amount of cosmic dust that arrives on earth is huge: it totals tens of thousands of tons each year. Most of it burns up or vaporizes—hence the brilliant shooting stars—but a small fraction, perhaps 10 percent or so, reaches the surface as solid particles like the spherules found on the *Challenger* expedition. The *Boys' Life* article contained suggestions about how to collect some of this cosmic dust, and of course I tried to do so. I laid out sheets of white paper in our backyard, on the flat roof of a garage, and in a few other places as well: my personal cosmic dust collectors. And I did collect a lot of dust. But even though a large amount of extraterrestrial material falls to the earth every day, the probability of true cosmic dust particles landing on my few square feet of paper was small. That was especially true for large grains, because much of the cosmic dust that makes it to the earth's surface is so small that the particles can only be seen with the aid of a high-powered microscope. Besides, my collectors were not left out for long, and we lived downwind of an industrial city that boasted two large steel companies. Most of the material that landed on my sheets of paper probably came straight from their blast-furnace chimneys. Still, none of this spoiled the excitement of the experiment. And because of it I never forgot about cosmic dust.

Although my experiment was unsuccessful, extraterrestrial material is ubiquitous on the earth, and if you know where to look you can find it relatively easily. This is especially true if you search in places where the influx is integrated over long periods of time, such as slowly accumulating deep sea sediments or Antarctic ice. Most of us do not have access to such materials, however, but the work of an inquisitive man named Jon Larsen has shown that, with a little diligence, anyone can collect cosmic dust particles without access to special instruments or samples, and without traveling far from home.

Larsen is a jazz musician from Norway. One day in 2009—it was summer, and he was outside, eating strawberries—a tiny gleaming particle on the table caught his eye. It sparkled in the sun and had a rough feel. It sparked his curiosity. Where did it come from? Might it be from space? Larsen was determined to find out. It took him a while: he had to read scientific papers, talk to scientists who studied cosmic dust, and learn how to distinguish true cosmic grains from the overwhelming preponderance of earthly particles. But he did find extraterrestrial particles, many of them. It was, he explained, like learning a new language. He began to look at ordinary dust in a new way, and it reminded him of the words of the musician Frank Zappa: the mind is like a parachute; it only works when it's open. Through his Project Stardust, dedicated to teaching amateurs about the extraterrestrial particles that are constantly drifting down though the atmosphere and settling on the earth's surface, Larsen has stimulated worldwide interest in cosmic dust collection by citizen scientists. He has also published a book with stunning photographs of some of the extraterrestrial grains he has collected, mostly by sifting through the dirt and debris that accumulate on the roofs of buildings. According to Don Brownlee, a professor of astronomy at the University of Washington who has studied cosmic dust for much of his career, extraterrestrial material is all around us: we breathe it daily and ingest it when we eat our vegetables. That gunk in the roof gutter you've been meaning to clean out? It may have a few small spherules from outer space mixed in with it.

The *Challenger* spherules, and the particles Jon Larsen found on rooftops, had escaped vaporization during their passage through the atmosphere. Survival depends on size: particles in a particular range—a "Goldilocks zone," not too big and not too small—heat up and melt but do not entirely vaporize. The tiny melted globules cool off quickly and freeze into spherules, typically less than a millimeter (0.039 inches) in diameter. Technically they are "micrometeorites"; today's investigators reserve the term "cosmic dust" for even smaller particles. But the *Challenger* scientists called their spherules cosmic dust, and for simplicity I use the term for all the small particles that land on the earth, regardless of size.

The question of why there should be cosmic spherules on the seafloor was not the only aspect of these objects that puzzled the *Challenger* scientists; another concerned their location. The scientists did not find the spherules everywhere; they seemed to be almost exclusively present in a type of ocean sediment they called "red clay"—something else that no one knew existed on the seafloor until it was discovered during the expedition. Red clay is found only in the deepest, most remote parts of the ocean, far from land, where very little material from the continents reaches and most of the biological debris that slowly drifts down from the overlying surface water dissolves long before it arrives at the seafloor. As a result, in such places the steady rain of extraterrestrial particles is not heavily diluted by other materials, and more spherules can be found in a given amount of sediment than anywhere else.

Most of the spherules contain iron metal and are magnetic, which makes them relatively easy to separate from the other components of deep sea mud. If a large quantity of red clay is stirred in water to make a soupy suspension and the slurry passed through a magnetic field, the cosmic spherules will be retained while the rest flows through. This is an efficient way to collect tiny particles from outer space. I have watched colleagues process sediments in this way, and I can attest that it is much more effective than putting out pieces of white paper in my backyard. Of course, obtaining a large sample of red clay in the first place requires a full-fledged seagoing expedition, which is not something that is easily accessible to an amateur cosmic dust collector.

In the search for cosmic material, some scientists have gone a step farther. Instead of bringing sediment samples back to the laboratory for processing, they have taken the magnets to sea. Fashioned into instruments irreverently referred to as "cosmic muck rakes," these seagoing magnets have been dragged across the ocean floor in red clay areas, collecting many thousands of extraterrestrial grains.

But on *Challenger*, there were no cosmic muck rakes. When the expedition began, almost nothing was known about the nature of the deep ocean floor; the scientists on board had no inkling they would find red clay or cosmic dust. Their primary tool for discovering what was beneath them was the dredge, essentially a large mesh bag that was dragged along the seafloor and scooped up whatever happened to be on the bottom. The *Challenger* scientists were not the first to use dredges, but even so their anticipation and the suspense as the

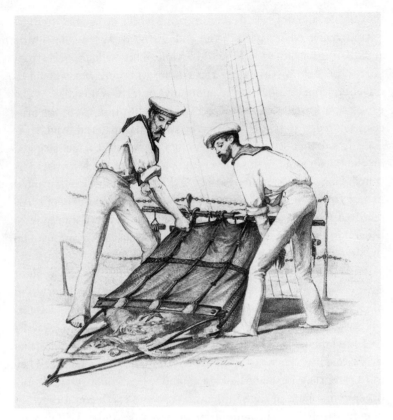

Sailors emptying the contents of a dredge onto the deck of *Challenger*. (Drawing by Elizabeth Gulland, courtesy of the Centre for Research Collections, University of Edinburgh.)

dredges were brought up—especially the first few that reached great depths—must have been tremendous. In the early days of the voyage, almost everyone on board, including hardened and indifferent sailors, was drawn to the deck as each dredge haul was brought in. They all wanted to see what strange creatures or objects had been scraped up from the seafloor.

Dredging was a long and tedious process. In the deeper parts of the ocean it could take a whole day to lower the dredge, drag it along the bottom, and slowly and carefully bring it back up again. Often the operation started at daybreak and ended after dark, the ship rolling and drifting with the wind the whole time. This made progress from port to port slow, frustrating the sailors, who were accustomed to traveling from one location to another as quickly as possible. As the expedition wore on, many among the crew lost interest in the process, as dredge after dredge brought up similar material. And occasionally a dredge would come up empty or, worse still, get snagged on the bottom and be lost. Even the patience of the scientists was tested. Henry Moseley, one of the *Challenger* naturalists, commented in his journal: "It is possible even for a naturalist to get weary of deep-sea dredging." A ship's officer who, like Moseley, eventually published his journal as a book about the voyage, succinctly summed up the dredging operation in a single word: "drudging."

The *Challenger* scientists never knew what surprises might await them in a dredge, however, and this kept anticipation high. John Murray, another of the expedition's naturalists, had primary responsibility for examining the geological contents of the dredges—the rocks and mud. It was he who first noticed the strange spherules, and they vexed him. He had never seen anything like them before. They were rare, they were not biological, and they did not resemble any other component of the sediments. What on earth could they be? Eventually, after he had examined many spherules and had eliminated, one by one, various other possibilities, he came to the conclusion that they might be extraterrestrial.

What Murray's shipmates thought about this novel idea is not recorded. But they must have discussed it. The naturalists and the ship's officers were messmates, eating together daily, and meals were a time for exchanging ideas and information. The spherules were too small for the chemist, John Buchanan, to analyze; they could only be examined visually. Probably Murray took his scientific colleagues into the laboratory to view them through his microscope. Circumstantial evidence indicates that at least one of the scientists, Henry Moseley, was not convinced that the spherules were extraterrestrial.

Two of the cosmic spherules dredged up on the *Challenger* expedition. The spherule on the left, about 0.039 inches (one millimeter) in diameter, was dredged from 11,400 feet in the South Atlantic. The spherule on the right came from 14,100 feet in the South Pacific; it is about half the size of the one on the left. (John Murray and Rev. A. F. Renard, *Report on Deep-Sea Deposits Based on the Specimens Collected During the Voyage of H.M.S. Challenger in the Years 1872 to 1876* [London: Her Majesty's Stationery Office, 1891], plate 23.)

Colleagues said that Moseley rarely embraced the opinions of others until he had independently satisfied himself about their validity, even when the opinions came from figures of authority. In the popular book about the expedition he published in 1879, he does not once mention the extraterrestrial particles. And even years later, in 1881, his skepticism was clear when he came across a reference to the presence of cosmic dust in soil, in a book by Charles Darwin. Moseley immediately wrote to the famous naturalist saying that he could not credit the existence of "meteoric dust." He enclosed a paper by a German colleague, who also questioned the reality of cosmic dust, to back up his views.

In the meantime, however, Murray had pursued the topic vigorously. He first communicated his discovery to the Royal Society of

Edinburgh in 1876, only a few months after the epic voyage of *Challenger* came to an end. He confessed that when he first observed the spherules he had found it difficult to account for them. Perhaps Moseley's skepticism had colored his thinking. But once he had concluded that they might be extraterrestrial he set about looking for ways to test his theory. Essentially, he had to eliminate all other possible origins. Iron metal, present in most of the spherules, does not occur naturally in rocks of the earth's crust but is widespread in manufactured objects, and *Challenger* was full of equipment made of iron. Could that somehow explain the presence of the magnetic spherules? Were they actually small pieces of the dredge or some other shipboard equipment? There were many other possible sources of fine iron particles, too: industrial furnaces, locomotives, other ships. Small grains could be carried over long distances by wind and ocean currents. Could he rule out these possibilities?

In some of the same dredge hauls that contained cosmic spherules the *Challenger* scientists also found round, golf-ball to baseball–size objects that were rich in the element manganese. These too were something new, and the researchers dubbed them manganese nodules. How the nodules formed was unknown, but when cut open they invariably showed concentric rings in their interiors, similar to the growth rings of a tree. This implied that the nodules grew in place on the seafloor, very slowly laying down layer after layer of manganese. If a cosmic dust particle fell onto one of the concretions, Murray reasoned, it would be incorporated into the nodule and preserved. Crucially, it would be protected from external contamination. So he selected several of the manganese concretions, rigorously cleaned their surfaces, and—being careful not to use an iron hammer and keeping anything made of iron well away—broke them apart by simply banging them together. He could be confident that the interiors were pristine. Sure enough, inside some of the nodules he found magnetic spherules exactly like the ones he had separated directly from the deep sea mud. They were not contamination; they had to be natural.

In his quest to understand the cosmic spherules, Murray studied samples of the one kind of undeniably extraterrestrial material he had access to: meteorites. He found that certain types of meteorites shared characteristics with his spherules, and at the Royal Society meeting where he announced his discovery he set up a microscope so attendees could observe the similarities for themselves. Still, he remained cautious. In his written article he notes, "There are many minute particles of native iron in deposits far from land; . . . some of these particles are little spherules; . . . these last, as well as some other spherules which are magnetic, have *probably* a cosmic origin." The italics are mine. Murray, in a sentiment recognizable to all scientists, ended his paper by acknowledging that his conclusions were subject to the results of ongoing investigations. More research was required.

To continue his investigations Murray enlisted the assistance of a Belgian geologist named Alphonse Renard, who would eventually be his co-author on the *Deep-Sea Deposits* volume of the Challenger Report. With a few more years of research under their belts the two investigators were in no doubt about the origin of the spherules: in 1884 they published a paper in which the words "cosmic dust" figured prominently in the title. In it they describe in detail the microscopic characteristics of the spherules and document their distribution throughout the world's oceans. In the same paper and in similar detail Murray and Renard note the ubiquitous presence of another component of the *Challenger* deep-sea sediment samples: volcanic ash. The paper is now considered one of the classics of oceanography and is still cited today.

A fortuitous event—fortuitous for Murray and Renard, if not for many residents of Indonesia—lent additional authority to their work. In late August 1883, a volcano on the Indonesian island of Krakatoa erupted violently, with devastating consequences. Most of the island was destroyed; tens of thousands of people died, the majority as a result of the catastrophic tsunamis triggered by the eruption. The magnitude of the volcanic blast has been estimated to be many times that of the most powerful nuclear explosion unleashed by humans. But for Murray and Renard two observations were especially important. The first was that flotillas of floating pumice from the eruption were carried thousands of miles from Krakatoa. The well-traveled pumice occurred mostly as rounded lumps, shaped by collisions and abrasion as wind and waves jostled the fragments against one another. During the *Challenger* expedition, dredge hauls had occasionally brought up rounded, waterlogged pieces of pumice from the deep sea floor. This had puzzled Murray and the other scientists on board because the pumice lumps were often found far from any known volcano. But the floating pumice rafts from Krakatoa showed that such material could travel very far from its source. Puzzle solved.

The second Krakatoa observation was that fine volcanic ash thrown into the atmosphere by the eruption dispersed rapidly over great distances; there were reports that it reached North America and even Europe. Clearly some of this airborne material would eventually settle on the ocean floor, far from the volcano itself. Murray had identified volcanic particles in many of the sediment samples from *Challenger* dredge hauls, but—as with the pumice fragments—he had been puzzled at finding them hundreds or even thousands of miles from the nearest active volcano. Now he had an answer. The small particles had been lofted high into the atmosphere by an erupting volcano and carried by the winds, just as had happened to the Krakatoa ash. All large eruptions would have similar effects; it should not be surprising to find volcanic ash in *any* ocean sediments, regardless of their distance from active volcanoes. And Murray made a further observation as well: none of the Krakatoa ash resembled the magnetic spherules. Another possible hypothesis for the origin of the spherules—that they were volcanic—could be ruled out.

In the months after the Krakatoa eruption people around the globe were treated to spectacular sunsets. It was generally agreed that the phenomenon was caused by dust in the atmosphere, but opinions varied about where the dust came from. Most people who thought about the problem concluded that fine volcanic dust from the eruption caused the colorful displays. It was a simple matter of cause and effect. But a few argued that the brilliant sunsets were at least partly due to something else: tiny particles from space.

Their reasoning was based on an atmospheric phenomenon that had been known for millennia: the "zodiacal light." *Zodiac:* the word is derived from ancient Greek and its original meaning was "circle of little animals." The so-called signs of the zodiac correspond to groups of stars which, with a little imagination, appear to outline animals or animal-like creatures in the night sky—a lion, a bull, a scorpion—and as the earth moves through its annual orbit around the sun, each constellation appears to rotate into alignment with the sun's position as it rises or sets. The zodiacal light is similarly connected with the sun's position. If you can find a dark place away from the lights of a city, you can observe it for yourself shortly after sunset or just before sunrise. It appears as a faint whitish glow that has a roughly triangular form, the base of the triangle centered approximately at the place on the horizon where the sun has just set or is about to rise. It is more easily observable at certain times of year: spring and fall are best, although even at those seasons it is faint. A full moon can render it all but invisible. But in the days before widespread light pollution obscured the beauty of the night sky, it must have been clear to everyone. The first detailed written accounts of the zodiacal light appeared in the seventeenth century, and even at that time some of those who described the phenomenon ascribed it, correctly, to scattering of the sun's light by . . . cosmic dust. However, the tiny particles that cause this phenomenon are not found in the atmosphere; they are in space, orbiting the sun along the same plane as the earth. So we should not be surprised that at the time of the *Challenger* expedition some scientists thought that a portion of the dust in the atmosphere—maybe even most of it—was extraterrestrial. If space around the sun was filled with dust, the earth's gravitational field would pull some of it in, and it would end up in the atmosphere.

In their paper about the *Challenger* deep sea sediments, Murray and Renard debunked this idea, or at least they debunked the possibility that the dramatic sunsets observed after the Krakatoa eruption were due to extraterrestrial dust. Although the cosmic dust spherules they

The zodiacal light, observed February 20, 1876.
(From a print by E. L. Trouvelot. Courtesy of the
New York Public Library Digital Collections.)

separated from the *Challenger* sediments were proof that extraterrestrial material enters the atmosphere, the particles were very different from the fine Krakatoa ash. The timing of the colorful sunsets and the certainty of volcanic dust in the atmosphere linked the two. There might be *some* extraterrestrial dust as well, but it would be a minor component.

For a long time after the *Challenger* scientists discovered the spherules, little attention was paid to cosmic dust. Those interested in extraterrestrial materials found meteorites that had crashed through the atmosphere and landed on the earth's surface more appealing objects of study because they were large enough to be sampled and analyzed with existing techniques. Although relatively rare, meteorites range from a few ounces to hundreds of pounds in size, and most museums were willing to part with small amounts of their larger specimens for research purposes. But in recent decades the development of ever more sophisticated techniques for analyzing extremely small samples, coupled with new and improved collection methods, has reinvigorated the field of cosmic dust research. Deep sea sediments have continued to be a rich hunting ground, but so too is Antarctic ice, which accumulates slowly and integrates the dust influx over long time periods. Balloons and high-flying aircraft now also routinely collect dust particles directly from the atmosphere; at the highest altitudes a significant fraction of the dust is extraterrestrial. Ingenious sampling devices have been constructed to collect tiny particles in space from satellites and space missions. Research on this abundance of samples, many so small that they have to be studied with electron microscopes, has given scientists new insights into the kinds of materials that exist in our solar system beyond the earth, and even in interstellar space. It has also provided clues to conditions during the earliest days of the solar system, and to the kinds of materials that went into the making of the planets, including our own. The research has even provided hints that cosmic dust might have seeded the earth with the organic precursors of life.

Such profound insight into the nature of the universe and our planet is not a bad legacy of the pioneering discovery by the scientists of the *Challenger* expedition. It is worth remembering that there was no mention of cosmic dust in the planning proposals the Royal Society sent to the Admiralty in preparation for the voyage; its presence on the seafloor was entirely unexpected. The discovery of extraterrestrial spherules by the *Challenger* scientists offers a wonderful example of how curiosity about the world around us can lead to new, often unanticipated, discoveries.

There is no question that curiosity was one of the traits, if not the primary one, that motivated the scientists aboard *Challenger*. For the most part, today we consider curiosity to be a good thing; it helps us learn and understand, and teachers are always keen to inspire curiosity in their students, hoping to enhance their assimilation of new information and concepts. Parents boast that their children are curious about everything. True, curiosity has another side—it killed the cat, after all. And those in power, be they politicians, religious leaders, dictators, or royalty, have sometimes feared curiosity. Curious people are more likely to question the status quo. The result has been banned books, jail terms or worse for scholars and journalists, and limitations on the dissemination of information. But in Victorian Britain, in the run-up to the *Challenger* expedition, the positive aspects of curiosity dominated. Especially among the rapidly expanding middle class there was an explosion of curiosity. Curiosity led to knowledge, and knowledge fueled progress. Progress meant better ways of doing things—travel by rail instead of by horse, say—and better living conditions for a larger portion of the population. That larger group was also being better educated, further stimulating their interest in and curiosity about the world.

So an obvious question is, Were the naturalists on board *Challenger* exceptionally curious people, or were they simply reflecting the tenor of their time? I suspect the answer is "both." In his book *Why?* the astrophysicist Mario Livio summarizes research that suggests we inherit many of our personality traits, such as curiosity, from our parents and more distant ancestors. To a significant degree these are part of our genetic makeup. I say "suggests" because it is much more difficult to be definitive in the social sciences than in the hard sciences, even with well-designed experiments. We can measure very precisely the amount of carbon dioxide in the atmosphere and how it changes from year to year, but the most elegant experiments on personality traits can so far give us only qualitative information about the importance of inherited versus environmental influences. With that in mind, and before we set off on *Challenger's* globe-encircling voyage, let us take a brief look at the expedition's six scientists and the time in which they lived—and what part curiosity played in their participation in this groundbreaking journey.

All the civilians on board *Challenger* were from families that were at least moderately well off—one was the son of a surgeon, another, of a clergyman, a third, of a high government official. All were well educated from an early age. The environmental influences of their upbringings, then, must have primed their curiosity about the world. And one of the characteristics of many highly curious people that Livio describes—a penchant for following their own interests rather than those imposed on them—is evident for several of the *Challenger* scientists. From all accounts, at various stages of their formal education at least three of them, John Murray, C. Wyville Thomson, and Henry Moseley, devoted more energy to their own investigations of the natural world than they did to their school studies. They did not neglect their formal education completely, but they valued their own explorations at least as highly.

C. Wyville Thomson

C. (Charles) Wyville Thomson, the scientific leader of the expedition, was born in Scotland in 1830; his father, a surgeon with the East India Company, was away from home for much of Thomson's youth, but even from afar his influence was felt. His expectations were high; his son was expected to excel at his studies. No comprehensive biography of Thomson (or, for that matter, of any of the other *Challenger* scientists) has been published, but the available information indicates that this pressure to succeed did not have any negative effects. Young Thomson was happy in school, and he did indeed excel. And although the pre-university curriculum of that time concentrated on the classics, often with little or no formal instruction in the sciences, at a young age he was already an inveterate collector and observer of nature. Weekends and holidays saw him roaming the countryside around the city of Edinburgh, exploring its biology and geology and bringing home samples of plants, animals, and rocks for further investigation. When eventually he entered the University of Edinburgh it was as a student of medicine—this, perhaps, at the urging of his father—but he soon decided to focus entirely on the natural sciences, giving up any idea of a medical career.

Even then, though, he followed his own path. An obituary for Thomson relates that while at university he asked his professor of botany, Professor Balfour, for a certificate acknowledging that he had completed the class. (Apparently university recordkeeping was much more casual in those days than it is today.) Balfour reportedly replied, "I will willingly testify to your knowledge of Botany, but I cannot certify that you attended my class." Thomson, it seems, had been absent from many of Balfour's lectures. He had found other ways to learn.

Although he left the University of Edinburgh without taking the examinations necessary to be granted a degree (this was not unusual at that time), Thomson was already highly respected and he was appointed, in quick succession, to academic positions at universities around the British Isles, in Aberdeen, Cork, and Belfast, where he taught and did research in botany, zoology, and geology. With these diverse experiences beneath his belt he was appointed professor of natural history at the University of Edinburgh in 1870, at the age of forty. At the time it was a prominent and highly influential position, probably the preeminent post of its kind in Britain.

As a "naturalist" Thomson had always had an interest in the oceans, and his early years in Edinburgh, and later in Aberdeen, Cork, and Belfast, afforded him plenty of opportunities to explore nearby coastlines. In hindsight, though, his growing curiosity about the deep sea—the ocean far beyond the shoreline—can probably be traced to his encounter with geology. It was a curiosity that would lead eventually to the *Challenger* expedition, and it began with, or at least was nudged by, his appointment to the chair of mineralogy and geology at Queen's College, Belfast, in 1854. Up to that point, Thomson's primary research interests had been in the biology of living things. But with his characteristic enthusiasm, he now dove into the study of paleontology. He quickly built up the collection of fossils in the college's museum, and his work on these, coupled with concurrent reports from other researchers about "primitive" organisms recovered from depths in the sea where some naturalists thought there should be no life at all, led him to a series of questions about the oceans that would occupy him for years to come, culminating in the *Challenger* expedition. Two of these were: To what depth in the ocean can life exist? and, Can we expect to find ancient organisms that we have so far encountered only as fossils still living in the deep ocean?

Thomson was well placed to investigate these questions. His wide expertise in botany and zoology, together with his studies of fossils, gave him the ability to identify similarities between present-day organisms and their possible fossil relatives, and to work out links that might exist with the living "primitive" forms being recovered from the oceans. He was, in a real sense, exploring the intricacies of biological evolution. This may not seem unusual to us today, but to put his investigations in context, consider that Charles Darwin's ideas on evolution had not yet been published. Thomson's work comparing fossil and oceanic organisms began in earnest after his move to Belfast in the 1850s; Darwin's *On the Origin of Species*, which upended the field of biology, was published near the end of 1859. Thomson, as an up-and-coming figure in the field of biology, may have known about Darwin's work (Darwin had reached his conclusions about evolution long before they were formally published), and, in addition, Thomson's work, unlike Darwin's, was restricted to a limited group of organisms. Still, I find it interesting that at this early date Thomson was questioning the prevailing idea that species are immutable. Perhaps his precocity had something to do with the intense spirit of investigation so characteristic of his era.

The idea of mounting a global deep sea expedition—what would become the voyage of *Challenger*—did not arise full-blown. Thomson's developing interest in the oceans led him and his scientific colleague W. B. Carpenter, who was at the time vice president of the Royal Society, to propose that the British navy allow them to use one of its vessels for local oceanic investigations in deep water off the coasts of Scotland and Ireland. The request was granted, and during the summers of 1868 through 1870, Thomson, Carpenter, and a few other colleagues mounted several short cruises on naval surveying vessels, focusing mainly on investigations into the extent and nature of life in the deep sea. On the face of it they were able to refute definitively the idea that the deep ocean was barren of life; their deepest dredge, at over two miles, recovered an abundance of invertebrate creatures. Still, such was the resistance to the idea that life could exist and even thrive in the complete darkness and high pressure of the deep sea that the question would remain one that occupied the scientists on the *Challenger* expedition several years later.

In spite of sometimes atrocious weather that limited the work they could do, the early expeditions were deemed highly successful, and their success undoubtedly helped give Thomson and Carpenter the confidence to propose their much more ambitious plans for *Challenger*. Thomson wrote a masterful popular book based on the results of their 1868 and 1869 expeditions titled *The Depths of the Sea*. It contains much technical information, but his enthusiasm for his subject shines through. He had a knack for describing complex ideas in ways that could be easily understood by nonspecialists; commenting on the great pressure in the deep sea, for example, and how it might affect life, he writes, "The conditions of pressure are certainly very extraordinary. At 2000 fathoms [about 2.3 miles] a man would bear upon his body a weight equal to twenty locomotive engines, each with a long goods train loaded with pig iron." And he describes how sponges attach themselves to the seafloor thus: "*Syalonema* sends right down through the soft mud a coiled whisp of strong spicules, each as thick as a knitting needle, which open out into a brush as the bed gets firmer, and fix the sponge in its place somewhat on the principle

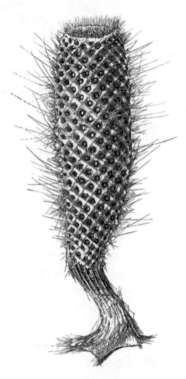

The sponge *Euplectella subarea*, a new species dredged early in the *Challenger* expedition and named by Wyville Thomson. The dredge, from a depth of about 6,500 feet in the Atlantic southwest of Gibraltar, brought up a diverse group of organisms including starfish, crustaceans, sponges, and sea cucumbers. This specimen was about a foot in length and two inches in diameter. (C. Wyville Thomson, *The Voyage of the "Challenger": The Atlantic*, 1:139.)

of a screw pile. A very singular sponge from deep water off the Loffoten Islands spreads into a thin circular cake, and adds to its surface by sending out a flat border of silky spicules, like a fringe of white floss-silk round a little yellow mat; and the lovely *Euplectella*, whose beauty is imbedded up to its fretted lid in the grey mud of the seas of the Philippines, is supported by a frill of spicules standing up round it like Queen Elizabeth's ruff."

Even the slightly stilted language of a nineteenth-century scientific report cannot hide the author's delight in the marvels of the natural world: "When one looks at the exquisite symmetry of these organisms [again he is writing about deep sea sponges], one almost wonders at the recklessness of beauty, which produces such structures to live and die forever invisible, in the mud and darkness of the abysses of the sea."

The *Challenger* expedition took Thomson away from Britain for three and half years; shortly after he returned he was knighted by Queen Victoria for his services to science. But although he went back to Edinburgh to resume his duties as professor natural history at the university, for the remainder of his career *Challenger*-related work occupied most of his time. He was appointed director of the Challenger Commission, an organization set up by the British government—at Thomson's behest—to ensure that the vast collections from the expedition would be properly catalogued and studied, and the findings written up and published. There was pressure to have the *Challenger* collections curated and studied by staff at the British Museum, the preeminent museum of natural history in the country, but Thomson argued that he and the other expedition scientists should have the privilege, at least initially, and he prevailed. As a result the Challenger Office was set up in Edinburgh.

For Thomson, managing the Challenger Office was a double-edged sword. It gave him ultimate authority to decide who should investigate the expedition's samples, but the administrative burden was so great that he had little time to carry out his own research. There were literally thousands of specimens and artifacts to deal with—everything from seafloor rocks and mud to fish, birds of paradise, plants and insects, even human bones and relics from remote oceanic islands. Sadly, Thomson never did finish the investigations he had hoped to make, nor did he live to see publication of the fifty volumes of the marvelous Challenger Report. He died, at age fifty-three, in 1882; the last volume of the report did not appear until thirteen years later, in 1895.

John Murray

Many people feared that the Challenger Report would never be completed without Thomson's guiding hand. Fortunately though, he had early on appointed John Murray as his second in command at the Challenger Office. Murray took on more and more of the responsibilities of running the office as Thomson's health deteriorated, and after Thomson's death he became director. His experience and abilities made him ideally suited to the task, and he and the Challenger Office became the nerve center for everything related to the expedition. Murray and a small group of colleagues ensured that key specimens were distributed to experts around the world for analysis; they received and often studied themselves samples from other scientists that might be relevant to the *Challenger* collections; and they prodded procrastinating authors to send in their manuscripts for the report. It was a monumental endeavor, and it is generally agreed that it only succeeded through the force of Murray's personality. He once claimed that he had personally proofread every word of every one of the fifty Challenger Report volumes.

Murray was only eleven years Thomson's junior, but he had barely begun his scientific career when he embarked on the *Challenger* expedition. Born in Canada, where he received his early education, he traveled to Scotland at the age of seventeen following the death of his father, with the idea that he would live with his Scottish grandparents and continue his education. (Both his parents were originally from Scotland; they had emigrated before his birth.) We know little about Murray's childhood in Canada, only that he was born and lived in Coburg, Ontario, then a regional center and important port city on the north shore of Lake Ontario. Perhaps Murray's lifelong fascination with ships started in the harbor at Coburg, although he himself later claimed that his interest in the oceans began with his voyage across the Atlantic to Scotland and his encounter with tides and the distinct smell of the sea when he visited the coast of his new homeland. It was so different from the shorelines of his youth on the Great Lakes.

John Macfarlane, the Scottish grandfather with whom Murray went to live, had a profound impact on Murray's life. A wealthy businessman, he was already in his seventies when the young Canadian arrived, but he was embarking on a project that was dear to his heart and would occupy him for the remainder of his life: to build a world-class museum of natural history in the small village of Bridge of Allan, where he lived. There were probably several motives behind his museum plans. Bridge of Allan was a spa town, and Murray's grandfather had commercial interests there. He saw the museum as a way to attract tourists and expand the area's population. But beyond that, he was a philanthropist, and he viewed his venture—in the prevailing spirit of mid-nineteenth century Britain—as a vehicle for education and self-improvement. His museum would not only house the natural history displays, it would also have a small library and a painting and sculpture gallery. It would be, he hoped, a place where people could come to learn about the world beyond their small corner of Scotland. The most important impetus for the project, however, was that Macfarlane had a personal, abiding interest in natural history. His newly arrived grandson shared that interest, and Macfarlane soon appointed him curator of the new museum even as Murray continued with his schooling. Given Murray's curiosity about the natural world—he fairly quickly stocked the museum with his own collection of specimens from the local environs—this was a shrewd move. Unknowable to either of them was that the position would give Murray administrative and organizational skills, as well as a background in natural history, that would serve as a bedrock for a distinguished career in science.

Macfarlane saw the position of curator as a lifetime career for his grandson. He probably also saw it as a way to secure the future of his museum and thus his own legacy. He even set down in his will, in painstaking detail, exactly how his grandson should carry out his curatorial duties—right down to the number of hours he should spend at work each day. But Murray had other ideas. In 1863 he convinced his grandfather to let him enroll at the University of Edinburgh, although the older man agreed only on condition that he study medicine. Murray did not like the course, however, and his stint as a university student was short. He was soon back at the museum as curator. But a few years later, in 1868, on the tenuous grounds of having once been a medical student, he joined a whaling ship as surgeon and embarked for the Arctic. How the crew fared with such an inexperienced doctor on board is not known. But what *is* known is that Murray kept the ship's meteorological records, made notes about the appearance of birds and other animals, and described organisms caught in tow nets. He also studied bottom mud brought up in soundings. He was a man of boundless curiosity, and his journey on the whaler was a foretaste of his career as an oceanographer.

How did this sometime student of medicine end up as one of the most respected scientists of his generation, earning honorary degrees, a knighthood, and accolades from around the world? By the time he returned from his seven-month adventure on the whaler his grandfather had died. In a sense, Murray was then a free man, no longer subject to his grandfather's plans for him and the museum. He was ready to engage with the wider world, and he returned to the University of Edinburgh—pointedly not as a medical student but so that he could take courses in subjects that interested him. Among other things, he worked in the laboratory of P. G. Tait, a distinguished mathematician and experimental physicist, gaining valuable practical experimental skills that he later used in designing and handling seagoing instruments. In a sign of his continuing interest in the oceans, in his free time Murray and a well-to-do friend from Bridge of Allan conducted their own natural history investigations on the friend's yacht, mostly in the sea near Edinburgh but also occasionally in more distant waters among the islands off Scotland's west coast.

Reportedly, Murray thought examinations were a waste of time. He could not see the point; if there was something that interested him, he would pursue it. As a result he, like Wyville Thomson, never did take the examinations required to graduate from the university with a degree. But with Murray as with Thomson, this omission did little to hinder his career. When Thomson came to choose naturalists for the *Challenger* expedition, Murray was on his list. He was still working in P. G. Tait's laboratory, and it is likely that Tait, who was a friend and colleague of Thomson's, recommended him. Beyond his understanding of natural history, Murray had seagoing experience, he had practical skills, and he had curatorial expertise from his years of managing his grandfather's museum. All of these would make his presence on the *Challenger* expedition especially valuable.

The rest, as they say, is history. Through his work during the expedition and especially because of his efforts shepherding the fifty volumes of the Challenger Report through the publication process, Murray's name is inextricably linked to that of *Challenger*. He is credited with coining the term *oceanography*, a fitting attribution, because oceanography is an interdisciplinary science, and of all the scientists aboard *Challenger* Murray was probably the most wide ranging. He did not have specific expertise in any particular aspect of biology or geology or physics or chemistry, as the others did, but he was widely educated across the sciences.

Murray's broad interests, specifically his curiosity about how coral reefs develop on tropical islands, even resulted in his becoming— quite serendipitously—a successful, and wealthy, businessman when he discovered phosphate deposits on Christmas Island in Indonesia and founded a company to mine them. Murray's interest in coral reefs was a direct result of his work on *Challenger*, and he once claimed that the taxes and fees levied on his mine and its products earned the British government more than the cost of the entire expedition, including publication of the report. As often happens, the practical consequences of pure research could not have been foreseen. Murray died in a car accident in 1914 at the age of seventy-three.

Henry Moseley

Moseley's path to a berth as a naturalist on *Challenger* was more conventional than John Murray's. But much like Murray (and Thomson), as a young man Moseley was filled with curiosity about the natural world. It was that intense interest that eventually led to his participation in the expedition.

Henry Nottidge Moseley was born in 1844. His father was a respected mathematician, who also had religious credentials (which was not unusual at that time): he was an ordained deacon and priest in the Church of England, and was a practicing cleric. Until shortly before Henry Moseley's birth he was also professor of "natural and experimental philosophy and astronomy" at King's College, London. Clearly the divisions between disciplines were fluid in the mid-nineteenth century. Also clearly, Henry Moseley was born into a family with a scientific bent (something that was to continue; Moseley's only son, also named Henry, became a distinguished experimental physicist, although his promising career was cut short by his death in the First World War).

Moseley's early education was at the elite Harrow School, a public school for boys founded in 1572 under a charter granted by Elizabeth I and still in operation in the twenty-first century. He was not an outstanding student—possibly because there was no instruction in what really interested him, the natural sciences. But like Murray and Thomson, he supplemented his formal education with his own investigations. His house master at Harrow, Mr. Rendall, would later observe: "I was able to give them [Moseley and a like-minded friend] a large room together for some time, a little detached from other boys, and this formed a regular laboratory—not always the sweetest—for experiments on plants, preservation of insects, etc. . . . I sometimes feared that I should have to interfere on sanitary grounds. But we managed to let things alone, and the future professor developed his powers there admirably."

After Harrow, Moseley entered Oxford intending to study mathematics as his father had, or possibly classics. The choice of subjects was probably at the urging of his family, but neither classics nor mathematics appealed to him, and he might well have dropped out had he not been rescued by a family friend who recognized Moseley's interest in the natural world and introduced him to George Rolleston, the Oxford professor of anatomy and physiology (the title does not fully do Rolleston justice; he *was* an expert in human anatomy, but his research and personal interests also covered subjects as diverse as archaeology, anthropology, zoology, and classical literature). Rolleston accepted Moseley into his circle of students, and the young man was suddenly catapulted into an environment in which he could follow his real interests. A few years later, in 1868, he graduated with distinction in natural science. For the next several years Moseley studied medicine, first in Vienna, then in London, and again on the Continent, in Leipzig. But in 1871 he was invited to join an expedition to observe a solar eclipse in Ceylon (today's Sri Lanka), an event that effectively marked the end of his medical career. Medical studies, subsequently abandoned, seem to be a common thread among several of the *Challenger* scientists: Thomson, Murray, and Moseley.

While in Ceylon, in addition to carrying out his observational duties for the eclipse, Moseley traveled around the country making observations of the flora and fauna and assembling a collection of specimens to take back to England. He was especially interested in . . . planarians. These usually inconspicuous but quite common creatures—they are flatworms—have many fascinating characteristics, not least their regenerative powers: cut up a flatworm and each piece will grow into a new worm, complete with head (including a rudimentary brain). In some circumstances they even grow multiple heads. For these reasons and others, planarians have attracted a lot of scientific attention. Moseley's curiosity about worms continued during the *Challenger* expedition. When the ship docked at Cape Town for repairs and outfitting he went looking for an unusual species of "velvet worm" that he knew was native to South Africa. After considerable searching he found one, and then holed up in a hotel for two weeks studying its anatomy while most of the others from the ship's crew busied themselves attending parties and social gatherings. Moseley was not antisocial, but no one had examined this peculiar worm in detail before, and he was determined to learn as much about it as he could. His efforts were rewarded. The worm, *Peripatus*, turned out to be a kind of "missing link" between worms and insects, having characteristics of both. Some of Moseley's contemporaries considered his work on this creature to be the most important biological discovery of the expedition.

But to return to Moseley's collection of planarians from the solar eclipse expedition: he took them back to England and began to examine them in the laboratory of his old professor, George Rolleston, at Oxford. However, his studies did not last long. Recognized as one of the country's promising young biologists—and probably recommended to the Royal Society's *Challenger* committee by Rolleston—he was soon appointed as a naturalist on the expedition.

From all accounts, Moseley was exceptionally well liked both by his fellow scientists on board *Challenger* and by the officers and crew. His dedication to the work at hand was legendary; no hardship seemed to put him off. One of the ship's officers wrote that Moseley was "chock-full of science, even unto bursting." He was especially interested in the flora, fauna, and—when present—human inhabitants of the remote oceanic islands visited by the ship. Usually he was the first ashore, and often he was the last to leave. Once, on the island of Kerguelen, he was unintentionally left behind as the ship sailed away. When his absence was discovered, his shipmates scanned the shoreline with field glasses: there was Moseley sitting by a rock, unperturbed, a makeshift flag (his handkerchief tied to a stick) stuck in the sand as a signal.

The *Challenger* expedition stimulated Moseley's interest in anthropology. When he encountered aboriginal people, especially those of remote islands in the Pacific, he worried that their cultures, their languages, and indeed the people themselves, were rapidly disappearing under a wave of European and American colonists. As a result he tried to study and record everything he could about them. Some years after he returned to England, he succeeded his mentor George Rolleston as professor of human and comparative anatomy at Oxford, a position that gave him the opportunity to continue his anthropological interests, and in addition to his own research he helped set up the Pitt Rivers Museum of Anthropology at the university. Moseley died in 1891, just before his forty-seventh birthday—his early death hastened, according to contemporary accounts, by overwork and, with his wide interests and responsibilities, the frenetic pace of his life.

Rudolf von Willemoes-Suhm

Rudolf von Willemoes-Suhm was the youngest of the civilians aboard *Challenger*; he had just turned twenty-five when the ship sailed. We know relatively little about his early life except that he was born in northern Germany and, like the other naturalists, showed an early interest in nature—he was especially fascinated by birds and had published papers and lectured about ornithology before the age of twenty. But when he went to university in Bonn it was to study law, following in the footsteps of his lawyer father. In a now familiar story, the subject that he initially chose to study was not one that interested him deeply, and within a year he had moved to Munich to study zoology. There he found his niche, a place where he could indulge his natural curiosity, and in quick succession he completed a Ph.D., did fieldwork in marine biology in the Mediterranean and the Baltic, and was appointed lecturer and assistant to one of Germany's top biologists at the University of Munich.

Willemoes-Suhm was a late addition to the scientific crew of the *Challenger* expedition. In 1872 he joined a Danish expedition to explore the natural history of the Faeroes Islands, and when the ship docked at Edinburgh to refuel, he took the opportunity to meet Wyville Thomson. He knew of Thomson and his dredging exploits by reputation, and evidently their meeting had been arranged in advance; in a letter written to his mother Willemoes-Suhm explained that the ship would call at Edinburgh, and "there I shall see the deep-sea fisher, Professor Wyville Thomson." Thomson may have had some inkling of the young naturalist's talents from the work he had already published, or perhaps through communication with his mentor in Munich, Professor Carl Theodor von Siebold. Whether or not Thomson already knew about Willemoes-Suhm's abilities, it is obvious that the young man impressed him deeply during their first meeting. Thomson recognized a kindred scientific spark in Willemoes-Suhm, and after just a short conversation and a few questions about his seagoing experiences, he asked him to join *Challenger* on its round-the-world expedition. The invitation must have come as a complete surprise to Willemoes-Suhm. The voyage was to get under way in a matter of just weeks, and the planning was already well advanced. Other members of the scientific party had been appointed months earlier. Nonetheless, Willemoes-Suhm was thrilled.

But before he could formally join the expedition, he had a hurdle to cross. Although Thomson was the scientific leader, he did not have unlimited authority and could not, or would not, add another naturalist to his small group without consultation. He needed support for the appointment, so he sent Willemoes-Suhm to London to meet with Thomas Huxley, an eminent scientist, the professor of natural history at the Royal School of Mines, and, more important, an avid supporter of the *Challenger* expedition and a member of the Royal Society's Circumnavigation Committee. If Huxley approved Willemoes-Suhm's appointment, there would be no problem.

Willemoes-Suhm's ship was scheduled to stay only a few days at Edinburgh, after which she was to sail down the coast and make a port call at London. So the young German scientist made arrangements to leave the Danish expedition temporarily, take the train to London for the meeting with Huxley, and rejoin the ship there. The meeting went well, and a few weeks later—by now the Danish expedition had returned to Copenhagen—he received a telegram officially inviting him to join the *Challenger* team as a naturalist. For Willemoes-Suhm it was a dream come true. "For the next three years," he wrote shortly after receiving the news, "I shall have the advantage of living in a house that moves slowly but surely through all five parts of the world."

But in one of the true tragedies of the expedition, he was destined never to return to his home in Germany or fulfill his considerable potential as a scientist. During the third year of the expedition, as the ship traversed the Pacific on her long homeward journey, he contracted a serious bacterial infection. Antibiotics were half a century in the future, and although the ship's doctors treated him as best they could, he died at sea. Willemoes-Suhm was well liked, and his death hit everyone aboard the ship hard, particularly his scientific colleagues. Wyville Thomson said of the young man's passing: "This sad event naturally threw a gloom over our little party. . . . I regarded Rudolf von Willemoes-Suhm as a young man of the highest promise; certain, had he lived, to have achieved a distinguished place in his profession; and I look upon his untimely death as a serious loss, not only to this expedition, in which he took so important a part, but also to the younger generation of scientific men, among whom he was steadily preparing himself to become a leader."

Henry Mosely also wrote about the impact of Willemoes-Suhm's death. In his journal Moseley usually concentrated on his own experiences and observations, only rarely mentioning his scientific colleagues, and even then typically in reference to a specific observation or publication. He seldom revealed any emotion. But of Willemoes-Suhm he wrote poignantly, "The voyage to Tahiti [from Honolulu] occupied a month. It was painfully impressed upon the memories of us all by the death of Von Willemoes Suhm. . . . I sat with him during the whole of the *Challenger* voyage, working day after day with the microscope at the same table."

What was Willemoes-Suhm like in person? We get a glimpse of the man beyond the dedicated scientist from the letters he wrote from the *Challenger*—mostly to his mother, but also to scientific colleagues in Germany. Here are a few snippets, in translation from the original German, that give us an inkling of his character, and also of what life was like on *Challenger*. The first dates from the beginning of the voyage, after a few weeks at sea. He and his fellow scientists had settled into a shipboard routine: "As to life on board, we have one meal after the other with complete regularity, and between them we smoke and work." A few months later they landed at Saint Paul's Rocks in the central Atlantic, and many on board took the opportunity to go fishing. Willemoes-Suhm spent his time collecting rock and mineral samples for colleagues in Munich. He observed his fishermen shipmates with amusement: "It is funny how much the English love to fish. They consider it capital fun, a great spree, and sit with infinite patience under the hottest sun exercising their sport."

After crisscrossing the Atlantic several times and being fitted out for cold weather in Cape Town, the ship sailed south toward the Antarctic. The weather steadily deteriorated. "It is so cold in the workroom that our fingers get stiff," Willemoes-Suhm wrote. "If it gets too cold, I bury myself under luxuriant sheep hides from Namaqualand (not far from the Cape of Good Hope), light lamps and candles, and to improve my mood read *Baron Hübner's Travels around the World.*" Soon after he wrote those words, the ship reached Heard Island, a bleak volcanic peak in the Southern Ocean. The island had been discovered only some twenty years earlier. "Some people in England . . . thought it would be a fine place for an astronomy station," he commented; "these people would be well advised to look at the island themselves—that would cure them once and for all." Much later in the voyage, with just over a year to go until he expected to be home again, the strain of long stretches at sea begins to peek through. In a letter to a colleague in Germany he muses, "I am often overwhelmed by a nameless longing for the passage back into the Elbe; this they call, I think, home-sickness, but the beloved animals [the marine creatures he was studying during the expedition] keep it from becoming too bad." Not long afterward the ship docked in Japan, and he wrote to his mother, "It is very strange; ahead of me is a carefree, beautiful day in a country I looked forward to, the most curious in the world, where I have never been before—and in spite of this, I am quite indifferent to it all just at this moment. I would rather be at the Parade Square at Rendsburg, go home, and have you tell me everything. The last time you heard from me was from Zamboanga; two months of rare monotony have passed since then . . . nothing but sea, water, and more water, no fresh food at all, and infernal heat."

The impression left by these brief paragraphs is of a sensitive young man with a keen eye for the world around him and a quiet sense of humor, dedicated to his work, but feeling the isolation and longing for the home he would never reach as the expedition pushed into its third year. After his death, Willemoes-Suhm's fellow scientists had

Wyville Thomson (left) and Rudolph von Willemoes-Suhm working together on *Challenger*. (Sketch by Elizabeth Gulland, courtesy of the Centre for Research Collections, University of Edinburgh.)

a memorial stone made up and sent to his family in Germany. The inscription reads:

IN MEMORY

OF

RUDOLPH VON WILLEMOES-SUHM

NATURALIST

WHO DIED ON THE 13TH OF SEPTEMBER 1875

AND WAS BURIED AT SEA

IN THE SOUTH PACIFIC OCEAN

ERECTED BY HIS MESSMATES ON BOARD

H.M.S. CHALLENGER

John Buchanan

John Young Buchanan was a chemist, which set him apart from the other scientists on the expedition, all of whom had backgrounds in biology and natural history. Born into an affluent family, he attended Glasgow University, where he first developed his interest in chemistry. It was a time of transition in Scottish universities, in which the physical sciences were beginning to be recognized as legitimate topics for study, and Buchanan embraced chemistry with a passion. After graduating he continued his studies at universities in Germany and France, combining chemistry with mineralogy. He quickly gained a reputation as a meticulous and practical experimentalist, and when he returned to Britain he became an assistant to the professor of chemistry—a post that had only recently been established—at the University of Edinburgh. Metaphorically if not literally Buchanan was working down the hall from Wyville Thomson, and when it was decided that *Challenger*'s scientific team needed a physical scientist, someone with experience in analytical chemistry who could also handle a range of instrumentation, he was an obvious choice.

Buchanan made several important discoveries during the expedition, as will be told later in this book. He was a perfectionist who strove to make his procedures as accurate as possible, and like the other scientists on board, he was rarely satisfied until he had pursued whatever question he was investigating to its conclusion. His published papers go to great lengths to describe every detail of the experimental methods he followed. One of his tasks on *Challenger* was to measure the density of every seawater sample collected—density being an important parameter for understanding how and why different water masses move in the ocean. The best way to ensure accurate density measurements, he wrote, was to make multiple observations on each sample using a hydrometer (an instrument based on the ancient principle that the amount of water displaced by a floating object is proportional to its density) and average the results. For each measurement he meticulously recorded the time and the temperature in his logbook; knowing the ambient temperature, he reminds us in his reports, is especially important for density measurements. Usually he made nine repeat measurements on each sample, and he notes that the process—all nine measurements—took him about twelve minutes. He does not explain why he chose nine, but by averaging the results he was able to reduce the uncertainty of his density determinations to small variations in the fifth decimal place, an impressive accomplishment.

Ships are not stable platforms, something Buchanan had to come to terms with as he worked in his small chemical laboratory. To counter the motion of *Challenger* when he was taking his density measurements he used a "swinging table" that introduced a modicum of stability—or at least reduced the effects of the more violent lurches of the ship. Still, when the sea was really rough the instruments were less important than the observer. His attention, Buchanan wrote, was "entirely taken up in looking after [my] stability, and in preventing [my] coming into collision with the swinging table." The concluding sentences of his discussion of seawater density in a paper he wrote long after the expedition wonderfully sum up his approach to work on *Challenger*: "It is, perhaps, not wholly unnecessary to point out that to obtain good results with a method such as this the observer must have a certain amount of dexterity and patience, but more particularly he must approach the matter with the desire to succeed. There is never any difficulty in making unsuccessful experiments."

Buchanan was twenty-eight when *Challenger* embarked on her round-the-world voyage. Independently wealthy, he did not have to worry about finding employment when he returned to Edinburgh at the end of the expedition. He set up his own laboratory and for a while worked on the samples and data he had accumulated during the voyage, and his papers and contributions to the Challenger Report earned him widespread recognition in scientific circles. In the following years he continued investigations into anything that interested him, much of it—undoubtedly motivated by his work on *Challenger*—involving work at sea. He sailed aboard ships laying telegraph cables, in his own yacht around Scotland, and on various oceanographic cruises with Prince Albert I of Monaco, with whom he became friends. In 1889 he was persuaded to take up a lectureship in geography at the University of Cambridge, and his inaugural lecture gives a clue to his widespread interests: it dealt with the effects of railways on trade, travel, and the opening up of new territories.

John Buchanan and his swinging table hydrometer.
(Sketch by Elizabeth Gulland, courtesy of the Centre for
Research Collections, University of Edinburgh.)

His talk was an enthusiastic endorsement of railways, but some years later he had second thoughts. Railways do have many benefits, he wrote in a subsequent paper, but he also noted that there was an emerging correlation between railways and the near extinction of many large land animals in both North America and Africa. That was not a good thing.

Buchanan gave up his Cambridge lectureship after only four years, but he so enjoyed the camaraderie and intellectual stimulation of the place that he continued to live in one of the university colleges for the next sixteen years. He published papers on global climate and its relationship to the distribution of land and sea, on the causes of earthquakes and volcanic eruptions, on the geometry of river mean-ders, on how to prevent train accidents and keep ships from sinking, and much else. In his private laboratories—in addition to the one in Edinburgh he eventually established a second in London—he un-dertook experimental studies simply to satisfy his own curiosity. Much of this work was known only to friends and colleagues; his fi-nancial situation meant that he was under no pressure to write up all of his investigations in the form of published papers.

Buchanan became increasingly pessimistic and depressed about the state of the world as World War I approached, and in 1914 he moved to Cuba, and later to Boston, to distance himself from Europe and the conflict. When he returned to Britain at the end of the war he lived mostly in London, gathering together his disparate publications into two large volumes that summarized his life's work. He took a few long sea voyages and maintained his connection with Prince Albert's Institut Océanographique in Monaco, but he was in the twi-light of his life. Buchanan died in London in 1925, the last surviving member of the *Challenger*'s scientific team.

John James Wild

Finally we come to John James Wild, the most enigmatic of the *Challenger* civilians. Relatively little information is available about him; most accounts identify him simply as Wyville Thomson's sec-retary or as the official artist of the expedition. His name rarely ap-pears in the books written about the expedition. But in addition to being a highly accomplished artist, Wild was a scientist in his own right. His best-known scientific work is a book published in London in 1877, a year after *Challenger* returned to Britain. Titled *Thalassa: An Essay on the Depth, Temperatures, and Currents of the Ocean*, it is largely based on data collected during the expedition, although Wild also incorporated whatever other reliable and relevant information he could find. The degree to which he was directly involved in the science observations on *Challenger* is unclear—he was, after all, Thomson's secretary, and when he was at sea his secretarial duties

occupied most of his time. But regardless of how actively he participated in the measurements themselves, Wild's book is a masterful
synthesis of a huge amount of data, most of it new, and he skillfully
lays out his own ideas about the implications of this information. A
few of his hypotheses, such as his suggestion that the great pressure
of seawater might be responsible for heating and eventually melting
the underlying crust of the seafloor, are fanciful. But others are not.
Throughout, he uses carefully drawn charts and graphs, most in
color, to help his readers visualize the data. His color-coded map of
depths in the Atlantic Ocean shows the shallow Mid-Atlantic Ridge
(as we now know it) that snakes down the center of the Atlantic almost as clearly as do present-day maps. Before *Challenger* the existence of the continuous ridge was unknown. His profiles of water
temperature versus depth in various parts of the ocean, and his discussion of the effects of temperature and salt content on density in
relation to ocean currents, presage current ideas about the so-called
thermohaline circulation. His maps of surface currents vividly illustrate the great gyres of the Atlantic and Pacific Oceans. Wild clearly
understood the wider value of the work carried out on *Challenger*. In
Thalassa he envisioned the tens of thousands of samples brought
back—both biological and physical—housed one day in a Challenger
museum. That, he assured readers, would be a tribute to the British
nation, which was "always ready to promote the cause of knowledge."
Without a doubt, he went on, for future generations the collection
would be of more interest than "all the trophies of war bought at the
price of general ruin."

Wild was born Jean Jacques Wild in Zurich, Switzerland, but at
some point he moved to Belfast, where he worked as a language
teacher. It is likely that he met Thomson there—recall that Thomson
was professor at Queen's College in Belfast from 1854 until 1870.
Thomson must have learned about Wild's abilities as an artist,
because Wild is credited on the drawings in Thomson's *Depths of the
Sea*, his popular account of the dredging expeditions he conducted
in the North Atlantic in 1868 and 1869. In the preface to the book,
Thomson thanked "my friend Mr. J.J. Wild" for his skill in produc-

ing the illustrations, and went on to say, "Every figure was with him a labour of love, and I almost envy him the gratification he must feel in the result."

In one of the surviving letters from the *Challenger* voyage, written shortly after the ship departed from England, a naval officer briefly described each of the civilian crew members. Wild, he remarked curtly, was dyspeptic and seldom to be seen. But the weather was atrocious, and Wild—along with even some experienced seamen—was almost certainly seasick. In his own memoir, Wild wrote that his recollection of those first few days and weeks was fragmentary. He had not yet acquired his sea legs. For someone unaccustomed to life at sea, those weeks must have been torture.

Other letters, however, paint a very different picture of Wild from the naval officer's early and not particularly complimentary description. They portray a man dedicated to his work yet capable of enjoying himself in port. They also tell of the close friendship between the Wild and Thomson families. In one letter, written by Wild to Thomson's wife, he assures her that her husband is in robust health and that the two of them—Wild and Thomson—are making great progress in writing up results as the voyage progresses. He inquires after Thomson's son Frank, and he thanks Mrs. Thomson for forwarding a package he has sent for his own wife. In another letter, from Thomson to his wife, Thomson asks her to give his love to Mrs. Wild and tell her that he is working her husband very hard. Thomson writes that Wild's duties leave him little time to himself when they are at sea—but he makes up for it when they are in port and "plays billiards and flirts with girls of all colors from morning to night." What Mrs. Wild made of this we shall probably never know.

Strangely, after the expedition Wild did not enjoy the level of success that the other *Challenger* scientists experienced. *Thalassus* was followed a year later by a more general book called *At Anchor*, which outlined his experiences during the expedition and was illustrated with his drawings. But no other publications followed, and in 1881 he moved to Melbourne, Australia. The reason is not clear. Perhaps he simply had fond memories of his visit there with *Challenger* and wanted to return. Or perhaps he left Scotland because in 1881 Wyville Thomson's health was deteriorating and he had to resign his position at the University of Edinburgh and step down as director of the Challenger Office. If Wild was still Thomson's secretary, he might well have found himself without a job. Whatever the reason, Wild went to Australia with no firm offer of a new position and apparently had difficulty finding employment. He worked variously as a lecturer in languages at the University of Melbourne, a secretary, and an illustrator. But his experiences on the *Challenger* expedition remained important. A lengthy 1891 article from the *Sydney Morning Herald*, describing Wild as "of the *Challenger* expedition," recounts the details of a lecture he gave to the Royal Geographical Society of Australia about a proposed expedition to the Antarctic. The article makes it clear that Wild was regarded as an important figure in Australian scientific circles and an expert on oceanic exploration. But a few years later, in 1896, a brief notice appeared in *Argus*, a Melbourne newspaper. In a short column titled "New Insolvents," Wild was listed as bankrupt. The reason was given as "retrenchment" in his position at the University of Melbourne, and diminishing work as an illustrator. His assets, according to the notice, totaled five pounds. (Although accurate comparisons are notoriously difficult to make, a reasonable estimate is that this would equate to around five thousand U.S. dollars today.) His liabilities were almost twenty times this amount. Wild died in Melbourne four years after this notice appeared, at age seventy-two.

In this chapter I have focused on the civilian scientists on board *Challenger*, but I don't want to leave the impression that they were the only people on the expedition with a strong sense of curiosity and a hunger for knowledge about the world around them. Many of the ship's officers had been selected from a roster of volunteers who presumably found the prospect of such an unconventional assignment exciting (in the peacetime navy, they may also have thought that such an unusual appointment could help advance their careers). Three of these officers eventually published their journals as popular books about the voyage. Although they sometimes poke fun at their "philosopher" shipmates, the officers' accounts show how keenly they viewed the scientific work and how fascinating they found the discoveries made during the expedition. They also write at length about their own experiences and observations ashore, describing the people, industries, farms, and wildlife they encountered. One of the officers, Thomas Tizard, the navigating lieutenant and senior surveying officer for the expedition, spent several years after the voyage working with John Murray to produce charts, diagrams, and hydrographic notes for the Challenger Report. And in 1992, Philip Rehbock, a science historian at the University of Hawaii, published and commented on a remarkable collection of letters and journal excerpts from a man who held a relatively low position in the ship's hierarchy: the steward's young assistant Joseph Matkin. Most of the material in Rehbock's book had only recently come to light, and it provides, as Rehbock points out, a unique "below the decks" perspective on the expedition.

Matkin's letters reveal a curiosity about the world and desire to learn every bit as strong as those of the scientists and officers. Like John Wild, during much of the voyage he was kept busy with shipboard duties, and many of his journal entries are about his observations during port calls, when he had the opportunity to explore strange environments and experience new cultures. But when he had time he also wrote about the ongoing discoveries made at sea. Once, when Thomson gave a lecture about the scientific aims of the voyage to the ship's company, Matkin sat down afterward and wrote out what he could remember of the presentation. He included his summary in a letter to his family, and noted that he found Thomson's talk "very interesting," and "a great success."

three SHAKEDOWN

George Campbell, naval sublieutenant, said it best. *Challenger* left England on smooth seas running before a light wind, but then "we got a heavy gale, which shook us all nicely down into our places; close-reefed topsails—ship rolling like mad—sleep at a minimum—scientifics sick—stand up meals—crockery smashing—perfect misery—attempted joviality, &c." The words are from a letter he wrote home shortly after the voyage began. Not only was Campbell a sublieutenant, he also had an inherited title: he was the son of a duke, and in the literature about *Challenger* he is often referred to as Lord George Campbell. His letters, not originally intended for publication, were later collected into a book titled *Log-Letters from "the Challenger"*—one of many personal accounts of the expedition. Letters were at the time of the expedition the primary form of communication, not only for those separated by vast distances but also for people living only a few miles apart. For those on board *Challenger*, at sea for months at a time and ashore only in far-flung foreign ports, letters were the only way to exchange information with their loved ones or learn about what was happening in the wider world. By the 1870s Britain had an efficient system of international mail built around the country's sea power. Because of the navy's global reach and the voluminous, world-spanning trade carried out by private British vessels, virtually every port visited by *Challenger* was frequented by British ships. They all carried mail to and from the homeland. Not many of the huge number of letters that must have come and gone from *Challenger* by such means have survived, but those that do offer a detailed glimpse into life on board. Campbell's letters are lively, perceptive, and full of wry humor. He claimed that he had not altered them for print; any editing was restricted to "rounding off only the most ear-breaking angularities."

The first few weeks of the expedition had always been intended as a shakedown, a time to test the equipment and sampling methods, and train the crew in various scientific procedures, but no one had anticipated the literal shakedown Campbell described so vividly. Among the scientists, Wyville Thomson had by far the most relevant seagoing experience: he had dredged and sounded in the North Atlantic in 1868 and 1869. Murray had made scientific measurements on the whaler when he was technically ship's surgeon, and Rudolph von Willemoes-Suhm had been on the Danish seagoing expedition shortly before joining *Challenger*, but the rest of the civilian crew was new to this kind of work. And with the exception of Thomson and his secretary J. J. Wild, they did not know one another well. Now they were thrown together in cramped quarters where they would have to coexist for the next three and a half years, working toward a common goal. Even Thomson, the veteran dredger, found working on *Challenger* different from his previous experiences. For one thing the ship was much bigger than those he had sailed on previously. Because she sat higher off the water, her motion in heavy seas was accentuated, posing problems for lowering dredges, sounding apparatus, and other equipment over the side.

Much of the work *Challenger* was expected to carry out was new to the naval officers, too. They had been selected for their abilities from a large pool of volunteers, but their experience did not guarantee success. They would be largely responsible for overseeing the practical side of the scientific operations: maneuvering the ship to ensure that she stayed on station when measurements were being made, overseeing the winch operator to make sure the dredge, thermometers, and sounding apparatus were safely lowered over the side and retrieved, and making some of the more routine measurements themselves. It would be a vastly different regimen from the one they were accustomed to. Only when the dredge or other equipment was back on deck would the scientists take over.

Challenger's first port of call after leaving Portsmouth was to be Lisbon, Portugal, followed by Gibraltar, Madeira, and finally Tenerife in the Canary Islands. This part of the voyage was meant to be the shakedown phase, a time to work out any kinks that arose in the procedures. Everyone on board knew that kinks were inevitable in such a large, new, and complex endeavor, and it was crucial to have everything running as smoothly as possible before they set off for more distant seas and lands. The plan was that *Challenger* would make soundings (depth measurements) and various other observations along the way, as well as begin the dredging operations, but this work would be preliminary to the main phase of the voyage. Any useful data gathered, or specimens collected, would become part of *Challenger*'s record, but only after she left Tenerife would the expedition begin in earnest, with a transit across the Atlantic to the Caribbean, stopping at regular intervals—typically every two hundred miles or so—for an official observing station. Such, at least, was the plan. But the storm that George Campbell wrote about was so severe that the first week was a complete write-off. Most of the scientists, and even some of the experienced sailors, were sick and struggled to get out of their berths. Lowering equipment over the side in such conditions was impossible. Doing anything useful in the laboratories was also out of the question. Science would have to wait.

By December 30, 1872, nine days after the ship left England, the weather had settled down enough for the men to try a sounding. It was successful; they got an accurate depth reading. But as they were hauling the sounding apparatus back on board the line broke, and they lost everything attached to it, including the thermometer that was meant to record the bottom-water temperature. Later that same day they tried dredging, but the seas were still high, and the first attempt was a failure; when they brought up the dredge it had been turned upside down and was completely empty. They tried again, with better luck. This time, to the delight of the naturalists, several creatures were captured in the dredge—not many, to be sure, but they offered the scientists their first glimpse of organisms from the deep ocean. In his journal Thomson wrote that the dredge was successful even though "the number of species procured was small." Others were less sanguine. Campbell, clearly unimpressed, noted dryly that the dredge contained a few starfish, a shrimp, and a fish.

A few days later they tried sounding and dredging again, and again they lost the sounding line and thermometer, and again the dredge came up empty. The day after that, determined to get it right, they attempted another dredge. This time it snagged on the bottom and the men could not free it; eventually the line broke and they lost the dredge. It seemed as though anything that could go wrong did go wrong, a discouraging start for the scientists. Campbell was philosophical. "Experience teaches," he declared, "and doubtless we shall soon become first-class dredgers." The following day the ship reached the quiet waters of the Tagus River, which led the crew to their first port of call, Lisbon harbor. The scientists had little to show for their initial attempts to investigate the ocean floor, and when they finally dropped anchor all aboard were relieved. It had been a harrowing start to the expedition.

When *Challenger* set sail little was known about the deep sea floor beyond the immediate coastal areas. But scientific curiosity about this unexplored part of the world was growing rapidly, in significant part stimulated by the development of submarine telegraph cables. The first successful cable, connecting England to continental Europe, had been laid across the English Channel in 1851, just twenty-one years before the expedition began. The first continuously functioning transatlantic cable was established in 1866. The telegraph companies were eager to understand conditions on the seafloor and how these would affect their cables. Was the bottom flat or irregular? Were there currents that could move the cables or bury them in mud? Plants that would grow on them or creatures that would eat through the casings? What was the temperature? The questions were endless, and even though submarine cables were not their primary motivation, the scientists on *Challenger* clearly recognized the importance of their work for this new form of global communication. Indeed, Wyville Thomson mentions undersea telegraph cables in the first paragraph of his report on the expedition, noting that they had directed the attention of "practical men" to the largely unknown deep sea. Britain was the hotbed of innovation in the industry, and the source of much of the early capital investment. Information about the ocean and its deep bottom was crucial for success. For the "practical men" it could provide commercial advantages.

How did the *Challenger* scientists plan to examine the seafloor, two, three, or more miles beneath them? We can't roll back the tape of time, but we can try to imagine their situation. Perhaps the most fundamental measurement they had to make was depth. But sonar, which sends sound waves through water to determine distances, had not yet been invented. For the crew of *Challenger* the measurements had to be done mechanically. This was relatively easy in shallow water; sailors had been doing it for centuries. Put something heavy on the end of a line and throw it overboard; when you feel it hit the bottom the amount of line paid out tells you the depth. But that simple procedure was not so simple in really deep water. To begin with, the sounding line had to hang vertically from the surface to the seafloor to ensure an accurate measurement. This meant that the ship had to be kept stationary, not an easy feat on a windy day with miles of sounding line to pay out. And in very deep water it could be difficult for sailors to sense the seafloor, because the weight of many thousands of feet of sounding line could itself drag the line down long after the weight on its end had reached bottom, resulting in a massive tangle of rope on the ocean floor. Some marine charts available at the time of the expedition showed occasional isolated regions in the oceans that appeared to be preposterously deep. Most sailors suspected that these measurements were erroneous and had been caused by just such problems.

The sounding or dredging line itself was something else the scientists had to worry about: What sort of material should it be made from? Today we would take for granted that a suitable material would either be already available or could be devised, but in the 1870s the choices were limited. The weights used for sounding were heavy, and so was a dredge full of seafloor mud or rocks. It was crucial for the expedition's success that the line could cope with the strain, or at least that breakages did not happen often. But no material that existed at the time was ideal. Multi-strand wire rope that could withstand high stress was not available; the only options were either single-strand piano wire or hemp rope. Piano wire had its proponents; it is relatively compact and easily stored, and it descends quickly to the bottom. It is also exceptionally strong—in a piano it has to perform under constant high tension, with continual pounding from the hammers—but making soundings in deep water is a different matter. With a heavy weight on the end, a sudden lurch of the ship could put enough stress on the line to break it, especially if it had bent or kinked during previous deployments. Hemp rope—which was the choice of the *Challenger* scientists for both depth sounding and dredging—had its own problems. Although it came in different diameters for different uses, it was heavy and took up a lot of space. It had to be dried and examined, inch by inch, after each use. *Challenger* left England with close to a hundred miles of hemp rope of various thicknesses coiled and stored in every available space, and even so the supply had to be topped up on multiple occasions during the voyage as rope was lost or became too worn to use.

Loss of a section of line did not worry the scientists as long as they had more in reserve, but what did concern them was loss of the attached instruments—the thermometers and water samplers. Later in his career John Buchanan, a fervent proponent of hemp rope in oceanographic studies, took part in voyages on which he had to use piano wire to lower his equipment. "I never attached a thermometer to the wire without feeling that I was guilty of a form of cruelty—cruelty to instruments," he wrote. On *Challenger* the crew used an ingenious device called an accumulator that, if it did not entirely prevent breakage, at least lessened the possibility by protecting the rope from sudden stress. It was essentially a flexible cylinder made up of two wooden disks (the two ends of the cylinder) attached to each other by numerous (exceedingly strong) india-rubber elastic bands. The device hung vertically with a pulley attached to the bottom disk; the dredging rope ran through the pulley. If the ship lurched suddenly in a large wave or gust of wind the rubber bands would stretch, taking up some of the stress on the line. The accumulator also acted as a gauge of tension on the line. If the rubber bands stretched toward their limit—or if one or two of them actually snapped—the sailors operating the winch could immediately stop reeling in the line or even reverse it temporarily to relieve the tension.

By the end of the *Challenger* expedition, accurate depth readings had been recorded at each of the 362 observing stations the ship occupied, and at more than 100 other locations that were not formal stations. In almost all cases a small sample of the bottom was collected at the same time, because the "sounding machine'—the weight at the end of the rope—was designed to multi-task. It was much more than a dead weight; a brass tube ran through it and projected out beyond the weight so that when the apparatus hit bottom, the tube would push down into the seafloor and fill with sediment. A simple one-way valve prevented the mud from washing out as the sounding line was hauled up.

The bottom samples retrieved in this way were useful for finding out what the seafloor was made of, but that was about all. They were not big enough to give the biologists a comprehensive sample of the flora and fauna that lived in and on the seafloor sediments, and they could not provide Murray with enough mud and rocks to thoroughly characterize the bottom materials. Adequate sampling of the seafloor required a dredge. In spite of the early failures—and as Campbell predicted—experience coupled with some minor modifications soon enabled the crew to become "first-class dredgers." The day after the ship left Lisbon, they lowered the dredge in water that was a little over half a mile deep. It came up full of gray mud, which the naturalists eagerly sifted through in search of biological specimens. Their labor was rewarded. They found many organisms, most of them species that Thomson and others had dredged up from the North Atlantic in previous years. Their findings seemed to support a theory that was then current: seafloor life was much the same no matter where it was sampled. (Of course the naturalists knew that many more dredge hauls from throughout the oceans would be needed to confirm the idea.) Two days out from Lisbon, the crew brought up another dredge haul with almost identical contents. But sifting through the mud in search of tiny creatures was laborious work. Once more,

Campbell captured the moment: "The mud! Ye gods, imagine a cart full of whitish mud, filled with minutest shells, poured all wet and sticky and slimy on to some clean planks, and then you may have some faint idea of how globigerina [tiny floating organisms whose shells make up most of the sediment] mud appears to us. In this the naturalists paddle and wade about . . ."

The naturalists would continue to paddle and wade about in the dredged up sediments throughout the expedition, but after years of finding similar creatures in one haul after another even some of the scientists became blasé. Thomson, however, never lost his enthusiasm. He was present every time a dredge was brought up on deck, night or day. As scientific leader of the expedition, he may have seen this as his duty. But I suspect it had more to do with his appreciation of the natural world, and the thrill he got from anticipating what new things might have been scooped up from the seafloor far below. His reports from the expedition often reflect his enthusiasm. Frequently they verge on the lyrical. One early dredge haul came up full of a species of long-stemmed sea fan that was brilliantly luminescent. In his journal Thomson imagined these organisms in their natural seafloor habitat: "Their immense number suggested a wonderful state of things beneath—animated cornfields waving gently in the slow tidal current and glowing with a soft diffused light, scintillating and sparkling on the slightest touch, and now and again breaking into long avenues of vivid light indicating the paths of fishes or other wandering denizens of their enchanted region."

The dredges that brought up these treasures were simple affairs, essentially large mesh bags attached to an oblong iron jaw. A weight was placed on the line a thousand feet or more ahead of the dredge to ensure that it stayed on the bottom, and the ship was allowed to drift with the wind while the dredge dragged along the seafloor. But the mesh bags were fine enough to retain even the "minutest shells" that Campbell complained of, with the result that they often came up filled with huge quantities of gloopy, slimy mud that had the consistency of wet cement. The little brass tubes on the sounding

The dredging and sounding arrangements aboard *Challenger*. Beyond the maze of ropes from the sails and masts, two accumulators (the cylindrical objects) are visible, the nearest attached to a dredging line, the second to a sounding line. (*Report on the Scientific Results of the Voyage of H.M.S. Challenger During the Years 1873–76 Under the Command of Captain George S. Nares, R.N., F.R.S., and the Late Captain Frank Tourle Thomson, R.N., Prepared Under the Superintendence of the Late Sir C. Wyville Thomson, Knt., F.R.S., &c., Regius Professor of Natural History in the University of Edinburgh, Director of the Civilian Scientific Staff on Board, and now of John Murray, One of the Naturalists of the Expedition: Narrative; Volume 1, First Part* [London: Her Majesty's Stationery Office, 1895], fig. 12.)

machines brought up too little sediment, the dredges too much. And for whatever reason many of the mud-filled dredges contained little in the way of living creatures. The scientists needed to find a better way, so they experimented with other types of samplers.

The device that seemed to work best was the trawl. Like the dredge, it consisted of a mesh bag suspended on the end of a rope, but it was bigger and its mouth was larger and more flexible than that of the dredge. As a result it scooped up a more varied collection of living creatures from the seafloor. In addition, it was made with coarser mesh so most of the fine mud washed through (a small section at the bottom of the trawl was made of much finer mesh in order to retain a modest amount of fine mud—Murray in particular wanted to be sure the trawl captured enough for his geological studies). Campbell was happy with this new arrangement. He no longer had to contend with mounds of mud on his freshly swabbed decks. This did not stop him from complaining, however. The creatures caught in the trawl, he thought, especially the fish, would nicely enhance the ship's cuisine. But the naturalists would have none of it. They wanted them for science.

Challenger was a sailing vessel, and most of her travel was done under sail. But at each of the designated observing stations the engines had to be fired up to keep the ship in a fixed position against the wind and currents while measurements were made. Sounding was always the first order of business because other investigations were dependent on knowledge of the water depth. If the sounding was successful the scientists obtained three pieces of information: the ocean's depth, the water temperature at the bottom (a single thermometer was always attached to the line just above the sounding apparatus), and the nature of the sediment (from the mud collected in the brass tube of the sounding apparatus). Once a sounding had been completed the next set of measurements would be attempted—typically a seawater temperature profile, which entailed lowering a line with multiple thermometers attached at different depths. Over the course of the expedition, nearly five thousand separate temperature readings were made. Partly in fear of losing too many instruments if the line broke, a single lowering never involved more than eight thermometers, and sometimes fewer. Thus several consecutive lowering and raising operations were necessary to obtain data from the surface to the bottom, and the deeper the seafloor, the more lowerings were required. A different line was used for water samples, and only one sampling bottle was attached at a time; the line was laboriously lowered to the desired depth, left there for a short time, and raised up again. The entire operation had to be repeated for each depth until all the required samples were obtained. Sometimes a special device was sent over the side to measure the speed and direction of currents at different depths. Dredging or trawling along the seafloor followed, now with the engines shut off and the ship drifting in the wind, dragging the dredge along the seafloor behind it. Separately, net tows were used to collect fish and other creatures from surface or near-surface waters. If the weather permitted, some of these operations—such as the net tows—were done from a small boat sent off on its own while the ship was on station. With so many different operations involved, occupying a station was a complex and time-consuming procedure, especially in deep water. For some of the sailors and officers who were responsible for making sure that everything ran smoothly—and who were used to sailing promptly from one assignment to another, not sitting around for days in mid-ocean—it was also supremely monotonous. The tedium of extended periods at sea, coupled with grim rations and uncomfortable living conditions for the ordinary sailors, led to more desertions from *Challenger* than were usual on naval ships. But from all existing accounts, most of those on board took a grin-and-bear-it attitude toward their unusual mission.

Once something came back on deck—thermometers, water bottles, the dredge, the sounding machine—the scientists took over. They had two specially built labs on the main deck, a novelty at the time, where they examined, measured, and curated the material that had been recovered. One was a small chemical laboratory where Buchanan held sway, analyzing water and sediment samples. The other was a larger natural history workroom for the naturalists. It was equipped with microscopes, bottles and jars for storing specimens,

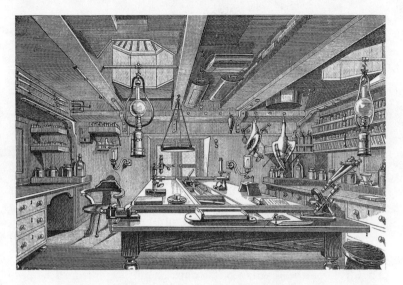

The natural history workroom on *Challenger*. John Buchanan's chemistry lab was much smaller. (*Report on the Scientific Results of the Voyage of H.M.S. Challenger . . . Narrative; Volume 1, First Part*, fig. 2)

and the various tools and instruments required for investigating the plants, animals, rocks, and sediments brought up from the deep.

Once again it is useful to imagine what conditions were like for these scientists, to put ourselves into their shoes and consider the procedures they followed. They spent a great deal of time and made enormous efforts to ensure that their measurements were as accurate and reproducible as possible. But the technology they had to work with was vastly different from that available today. The thermometers attached to the sounding line, for example, were crude by current standards. They were built to operate under the high pressures of the deep sea, and they were designed to register both a maximum and a minimum temperature; generally the minimum temperature was the one of interest because in most locations the deeper the thermometer went, the lower the water temperature. So the line with attached thermometers was lowered to a predetermined depth, held stationary until the thermometers had equilibrated with the surrounding seawater, and then hauled back up; the minimum temperature reading was assumed to be the water temperature at that depth. The only way to check accuracy was to repeat the measurement.

There was one exception, however. If a dredge at the same location had been brought up quickly, the temperature of the mud it contained, especially if there was a lot of it, would not have changed significantly. The scientists could stick a regular thermometer into the sediment as a check on the reading from the bottom-water thermometer. Usually the sediment temperatures, and most of the repeat measurements, agreed with the original observations, giving the scientists confidence that the maximum-minimum thermometers were accurate.

The water sampling bottles were another crucial bit of equipment. They had to be designed in such a way that only water from a particular depth was sampled; it was important that no contamination occurred as the line was fed out or hauled in. How could the water samples be securely sealed in the bottle at the desired depth? A few varieties of water sampling instruments already existed, but Buchanan, who was responsible for chemically analyzing the water samples, designed his own: brass bottles with springs and flaps that kept them open and flushed as they descended, and then shut tightly when the line was pulled up. He had spent many hours testing the sampling device before the expedition started, filling his bottles with dye, lowering them into the water and opening the flaps, and then sending them down to various depths to collect test samples. In water taken from depths beyond about ten feet he found no residual dye. That gave him confidence that the bottles worked as intended.

Indoor instruments were as important for the expedition's success as the dredges, thermometers, and water bottles. In the natural history laboratory, microscopes were essential pieces of equipment and in constant use. They were a window into the world of the tiny things collected from the sea: minute organisms from the water or the sediments, particles of cosmic dust, microscopic shells, and individual mineral grains from the bottom mud. By the 1870s, the optics of commercially available microscopes were excellent. But illumination was another matter. The ship had no electricity. During the day the scientists relied on the natural light that filtered in through the ports, and after dark they had to use candles or small oil lamps fitted onto the microscopes. Compounding the difficulty was the fact that the ship was always in motion, even in calm seas. The illumination problem, of course, was not restricted to microscopy; it extended to all work on the ship—especially work done below decks, where no natural light was available at all, even in daytime.

The first few weeks of the expedition turned out to be essential for helping the scientists determine which types of equipment were most effective, what procedures produced the best results, and who should be responsible for what. It was a learning phase for everyone, and although it continued after the shakedown period was over, by the time the ship left Tenerife to begin the expedition proper, the scientists, officers, and crew had settled into a more or less optimal routine. Beyond the overarching but vague goal of increasing knowledge about the ocean, however, what were Thomson and his small group of scientists hoping to learn?

The original proposal promised investigations of everything from ocean currents to the nature of the seafloor and the botany of remote oceanic islands. But Thomson and Carpenter, the principle proponents of the voyage, were both biologists, and their main interest was marine biology, especially the biology of the deep ocean. Most of the fifty Challenger Report volumes deal with biology. Thousands of new species of plants and animals are reported and described. The topography of the seafloor, its geology and mineralogy, the direction and strength of currents, the temperature and chemical properties of seawater at various depths—all of these are also discussed, and in some ways the global implications of these new observations were as important, or more so, than the biological findings. But in terms of sheer volume, biology won, hands down.

Thomson had several specific biological questions he hoped to address, as I outlined briefly in Chapter 1. One of these was whether life could exist at great depths in the oceans. His own dredging work in the North Atlantic had brought up organisms from depths of more than two miles. But he wondered about the even deeper parts of the oceans. Would he find organisms there too, or would these regions be devoid of life? On the basis of his earlier work Thomson had argued that life most likely flourished at all depths. Some scientists remained stubbornly unconvinced, however, believing that the extreme environment of the deep sea was inimical to living things. They suggested that the creatures Thomson and others thought came from depths of a mile or more could have been caught up in the dredges nearer the surface, as they were being hauled in. Thomson knew it would take overwhelming evidence to change their opinion; *Challenger*'s dredges from the deepest parts of the world's oceans, he believed, would provide it.

And he had further, related questions: What was the nature of life in the deep sea? Were living things more or less abundant there than on the surface? How did organisms that lived in the deep sea obtain their nourishment? Were they more primitive than creatures living at shallower depths? Would the deep sea be full of creatures that were known only as fossils from past geological eras? The lessons learned during the shakedown phase of the expedition gave Thomson and his fellow scientists confidence that *Challenger*'s ambitious sampling program would be able to answer these questions, and perhaps others they had not yet thought of.

Before we move from the shakedown period to the main expedition, I would like to touch on an additional aspect of *Challenger*'s voyage: visits ashore. Sometimes the ship's port stops were strictly for fueling and provisioning, sometimes for repairs, and sometimes she anchored at remote islands purely for science. Often stops and shore visits accomplished all three. During the shakedown phase, *Challenger* made four port stops: at Lisbon, Madeira, Gibraltar, and Tenerife. All these places were relatively familiar to the men on board; either they had visited them previously or they had heard about them from other seamen. For the scientists, the shore visits tended to be part busman's holiday, part tourist opportunity, and sometimes a chance to carry out on-shore natural history research. The civilian scientists were paid an annual salary by the Admiralty, but they were not officially naval employees and they were not subject to the navy's rules and regulations. For the most part they had few ship-related duties when in port, aside from self-imposed tasks such as examining samples in the laboratory or writing up their reports. In contrast to the naval personnel, they were free to come and go as they pleased and do as they liked.

When the ship made her first port stop at Lisbon, the voyage had barely begun—they had been at sea just under two weeks. The scientists, at least some of them, disembarked immediately. Thomson and a few others forsook their shipboard accommodation and took up residence in a comfortable hotel. They had, he noted, great views of the town and across the river. After being tossed about on *Challenger* for days they were happy to be again on terra firma.

Like most visitors to a new city one of the first things they did was go sightseeing. In his journal Thomson describes a visit to a famous monastery, the Mosteiro dos Jerónimos, on the western outskirts of Lisbon. (The monastery is still high on the agenda of most of today's visitors to Lisbon.) He includes sketches of the building and long descriptions of its intricate architecture. The monastery also engages his scientific curiosity: he describes the different rock types used in the buildings, the fossils he noticed in the building stones, the variety of plants in the cloister, and the movements of the hawk moths feeding from the flowers. By contrast, Moseley is all business. In his journal he never mentions Lisbon's cultural delights. We do not know whether he visited the monastery or, indeed, if he even left the ship. He does not begin his narrative of the expedition until later, in Tenerife, where he embarked on an excursion to collect specimens.

Thomson reports on other activities he engaged in during the Lisbon port call too, including visits to the Botanical Gardens, the Meteorological and Magnetic Observatory, and the laboratories and museums of a newly inaugurated technical school. He was impressed by the several museums attached to this school, particularly the display of birds in the zoological museum. The collection belonged to the king of Portugal, King Luiz I, who was interested in natural history and had a passion for oceans—he personally funded a number of marine research voyages. One evening he invited Thomson and *Challenger*'s captain, George Nares, to dinner at his palace. Later he visited the ship. The sailors grumbled about the extra cleaning and the preparations they had to make, but their complaints could not mask their pride in the ship and her mission. The king was not the only dignitary to visit *Challenger* during her stay in Lisbon. Among others, the British ambassador and his family came on board. These occasions—part social, part business—were a precursor to similar ship-to-shore exchanges that took place whenever *Challenger* anchored at foreign ports. Such was the interest in the voyage that local dignitaries often organized balls, dinners, and parties for the scientists and officers. In return, politicians, diplomats, businessmen, and local scientists were invited on board to see the array of scientific equipment on the ship and enjoy social events. Sometimes they were taken on short pleasure cruises. Not everyone on the ship enjoyed these social occasions. Herbert Swire, a sublieutenant, wrote in his journal about a ball held for the *Challenger* crew in Cape Town that "these balls are very dreadful to me and I only go 'on duty' and would just as soon spend the same time at the masthead or in the ship's cells." Much later, when the ship was in Tahiti, he escaped when "the Queen, Governor, and all the swells" came on board for similar festivities: "I am happy to say I was away that day, taking observations at Point Venus."

A little over three weeks after leaving Lisbon the ship arrived at Santa Cruz, Tenerife, where the naturalists organized the first of many onshore science excursions. This, too, was meant to be part of the shakedown phase of the voyage, a trial run for similar junkets that would follow in other parts of the world. The shore party included naval personnel as well as scientists, a policy initiated by Thomson to help give the crew a personal stake in the ship's scientific mission. The practice was followed throughout the expedition. For several days three of the naturalists, officer George Campbell, and two seamen explored Pico del Teide, the island's famous mountain, while *Challenger* took soundings and dredged around Tenerife and the other nearby Canary Islands. (There was also the obligatory social engagement: the British Consul organized a ball to honor *Challenger*'s visit. William Spry, an engineer on the ship, enjoyed these occasions immensely, unlike his shipmate Herbert Swire. When he arrived at the venue for the party, "all the available Spanish beauty" were there to meet them, he wrote. His only complaint was the difficulty of communicating with his dancing partners. But he noted that the Spanish women had cultivated the "language of the eyes" to perfection: "What the heart felt and the tongue could not utter the eye interpreted.")

Tenerife is a volcanic island and Mount Teide, its highest peak, is an active volcano that last erupted in 1908. (Based on its geological history it will erupt again some day, potentially catastrophically.) But the *Challenger* naturalists were mainly interested in Teide's natural history, not its geological past or future. At the time of their visit the mountain was already well known for its progression of vegetative zones, from palms and orchids at sea level to pine forests and, eventually, lichens and moss at its highest altitudes. The plan called for the scientific party to travel through these zones to the summit, collecting plants and animals along the way. On hearing this the locals burst into laughter. It was February, and the summit was still covered with snow; only the crazy British would try such a thing in winter.

At 12,198 feet, Teide is not only the highest peak on Tenerife, it is the highest point in all Spain. The little party from *Challenger* got up to about 9,000 feet, where they encountered snow and had to halt their ascent. They had hired guides, but these local men had already turned back, saying they would go no farther. During the day, even in February and at high altitude, the heat was relentless, and the scientists baked in the sun. But as evening fell the temperature plummeted. During the first night on the mountain the group's water supply froze solid. They burned dry broom bushes to keep warm, throwing flames and plumes of sparks into the air; the local villagers, seeing the fires from below, wondered whether the volcano was erupting.

The naturalists spent three days on the mountain, and they never did see the neat sequence of vegetation they had expected. Campbell was disappointed; the textbook illustrations of Mount Teide were all "snare and delusion," he declared. His dream of climbing through one well-defined zone after another had been "rudely dispelled." They saw very little in the way of living creatures, either: a few rabbits, a bird or two. Moseley found some half-hibernating centipedes, spiders, and beetles hiding under rocks. Campbell again: "In a natural history way our Teneriffe Peak cruise was rather a farce. We had been in hopes of getting plenty of insects and birds, but nowhere have I seen a greater absence of life in any shape." On the way down the mountain they were able to augment their natural history collection when they gave a local shepherd a shilling and he returned with a box full of snails and beetles.

The explorers from *Challenger* found only one item in abundance during their excursion: rocks. On much of the mountain there was little else, and they made a collection of various types, shapes, and sizes. But despite the meager scientific results, the trip was deemed a success. Logistically, in spite of the uncooperative guides, it had gone smoothly. Everyone had enjoyed himself. Moseley had been unable to collect as many biological specimens as he had hoped, but he wrote at length about the details of the sunsets and the ever-changing cloud formations that swirled around the mountain. His passages on Teide offer one of the few places in his journal where he deviates from his normally descriptive and analytical prose to express wonder and appreciation for the natural beauty of the scenery before him.

The naturalists went back to the ship confident that they would be able to mount successful exploration trips in the more remote regions *Challenger* was scheduled to visit later in the voyage. Valuable lessons had been learned both on land and at sea during the shakedown phase; now the expedition would begin in earnest. As the ship left Tenerife the sails were set for the long traverse across the ocean to the Caribbean. The next sounding and dredging site would be observing station number one, the first official station of the expedition.

four

**CONTINUOUS CHALK, EVOLUTION,
AND THE PRIMORDIAL OOZE**

We are all familiar with commercial chalk: the small pieces of soft material that can be used to draw pictures on a sidewalk or write on a blackboard. Natural chalk, though, can be a mile thick, and it sometimes forms stunning landscapes like the famous White Cliffs of Dover in southern England. If you broke off a piece of chalk from the White Cliffs, it too could be used to write on a blackboard or a slate. On a clear day the brilliant white rock faces can be seen from the other side of the English Channel in France.

That natural chalk is at one end of a spectrum of sedimentary rocks known collectively as limestone. The name limestone comes from the millennia-old practice of heating the rock to a high temperature, a process that drives off carbon dioxide and leaves behind calcium oxide (whose common name is lime), a key ingredient of cements and mortars. Before electricity, lime was burned to illuminate theater stages, bathing actors in limelight. Most limestone is made up of layer after layer of the cemented-together shells of marine organisms. It is a common rock type at the earth's surface; where I grew up it was the local bedrock. Chemically, limestone is made up predominantly of calcium carbonate, but it also commonly contains varying amounts of other things—clay, for example. Chalk, however, is as close to pure calcium carbonate as occurs in nature; it contains almost nothing else.

What does chalk have to do with the *Challenger* expedition? Well, to begin with, the *Challenger* naturalists were familiar with "the chalk," the pure limestone formations that make up the White Cliffs of Dover and outcrop across many other parts of southern England and western Europe. Geologists of the day recognized that these rocks had once been soft sedimentary layers on the seafloor that over time had become cemented into solid rock and thrust up above the sea's surface. As *Challenger* prepared to leave on her globe-encircling voyage, many, if not most, scientists believed she would find material much like the chalk everywhere on the ocean floor.

The European chalk formations are assigned to the Cretaceous period of geological history, a period that is defined by these deposits: *Cretaceous* comes from the Latin word for "chalk." In the 1870s neither the exact age of the Cretaceous nor its duration was known, but with modern methods of measuring time we now know that it lasted from about 145 million to 66 million years ago, and that the chalk of the White Cliffs and other similar deposits was formed toward the end of the period, between about 90 million and 66 million years ago. Dinosaurs still roamed the earth; it was a time of global warmth caused by high levels of carbon dioxide in the atmosphere, and conditions were ideal for the proliferation of the small marine organisms whose shells make up the chalk.

The first samples of seafloor mud ever recovered from deep parts of the ocean had been collected only a few decades before the *Challenger* expedition. Most were from the North Atlantic and had been retrieved in conjunction with sounding operations by naval vessels or ships preparing to lay undersea cables (like *Challenger*'s "sounding machines," other types of sounding weights sometimes brought up seafloor samples). The early bottom samples were small and few in number, and scientists were eager to examine them. When they were analyzed, it was quickly recognized that the seafloor sediment bore striking similarities to the chalk formations of southern England. Microscopic examination revealed that the tiny shells in the mud came from the same types of organisms that were found as fossils in the chalk. One of these was *Globigerina*, the variety of foraminifera that gives its name to the mud: *Globigerina* ooze. Wyville Thomson, who had collected seafloor samples of *Globigerina* ooze during his pre-*Challenger* voyages in the North Atlantic, was so impressed by the similarities between the chalk and the seafloor mud that in a popular talk in 1869 he proclaimed that one was just a continuation of the other. Everywhere he—and others too—had sampled Atlantic sediments they were the same; they were simply "modern chalk," not yet hardened into solid rock. A few other scientists had also commented on this, but Thomson went farther. Noting the uniformity of the deep Atlantic samples, he concluded that the *entirety* of the ocean floor must be a vast, never-ending landscape covered in a blanket of fresh chalk. Formations like the White Cliffs of Dover were simply bits of seafloor mud that had been elevated by some unknown process to form dry land. The vertical movements, he reasoned, were minor compared to the depth of the sea, and—by definition—they had only occurred around the edges of the ocean. Most of the ocean floor had always been submerged and must have been accumulating chalklike sediments at least since the Cretaceous. The last chapter of Thomson's *Depths of the Sea*, published just a few weeks before *Challenger* left England, was titled "Continuity of the Chalk." His emphasis was on continuity in time, but he also implied that there was continuity in space. His theory was about to be profoundly challenged.

The first surprise came during *Challenger*'s initial crossing of the Atlantic in the spring of 1873. The ship had just left Tenerife and was heading west toward the Caribbean. Close to the Canary Islands the dredge brought up volcanic rocks, presumed to be from the undersea ramparts of the islands. As *Challenger* moved farther west into deeper water, the dredge hauls were soon recovering pale, sticky *Globigerina* ooze, just as everyone had expected. The bottom of the deep Atlantic was indeed covered in a blanket of the ooze; the idea of "continuous chalk" seemed to be confirmed. But then a surprising thing happened. As the ship continued westward the ocean became deeper, and the character of the mud began to change. The first thing the scientists noticed was its appearance; it had a darker color, and when John Buchanan performed a chemical analysis he found that it contained considerably less calcium carbonate and more clay than the usual *Globigerina* ooze. The biologists examined the mud under their microscopes and discovered that some of the species they were accustomed to seeing—especially species with fragile calcareous shells—were missing. Most of the shells that were present were broken, corroded, and discolored.

Then, a little over 1,100 miles from Tenerife, in the deepest water they had yet encountered—the sounding read 3,150 fathoms, or about

3.6 miles—the dredge came up filled with more than a hundred pounds of something they had never seen before: smooth, reddish-brown and extremely fine mud. It was definitely not *Globigerina* ooze. The sediment particles were so fine that Campbell commented, "Between the fingers it feels like grease." When the mud was put in a jar of water it remained suspended for days, resembling, Thomson said, hot chocolate. Sifting through it, the *Challenger* naturalists found only a few tiny shells; there was little sign of anything else of biological origin. For Thomson this must have been a shock. He had been confident that the whole of the ocean floor was blanketed with the shells of *Globigerina* and other similar organisms, and here was something totally different. But the whole point of the expedition was to explore the deep ocean, and Thomson was happy enough to admit that he might have been wrong. The reddish colored sediment was a discovery, and it called for a celebration. According to the thermometer sent down with the sounding apparatus, the bottom-water temperature was 36.8 degrees Fahrenheit at this station. The red clay brought up in the dredge was the same temperature, a natural refrigerator, perfect for chilling champagne. Thomson got out a bottle and thrust it into the mud, and soon the scientists were toasting their discovery.

As *Challenger* continued sailing west, more deep-water dredges came up full of the reddish-brown mud. Clearly it was an extensive deposit. But then, about halfway across the Atlantic, the ocean shallowed. They were approaching what we now call the Mid-Atlantic Ridge, a feature that was then unknown. And as the depth decreased, the dredge hauls went through the same sequence the scientists had seen previously, now in reverse: the mud gradually became lighter in color and richer in calcareous shells. The foraminifera were cleaner looking, and most of the shells were intact and not corroded or discolored. Soon they were once again hauling up pure *Globigerina* ooze. But as they continued toward the Caribbean, the depth dropped off again, and the amount of biological material decreased dramatically. The dredges once more came up full of fine-grained, reddish-brown muck.

From our perspective it seems clear that these observations imply a correlation between water depth and the type of sediment on the ocean bottom. *Challenger*'s dredge hauls brought up reddish-brown mud only from the deepest seafloor; in shallower regions calcareous *Globigerina* ooze dominated. Wherever the water shallowed or deepened a gradual transition occurred between the two. But the *Challenger* scientists had no prior knowledge of what they would find. They were working out their ideas on the fly. The fine-grained reddish mud was entirely new; no one had ever seen it before. They were not even sure whether the shells they found in the calcareous ooze came from creatures that lived on the bottom, on the surface, or somewhere in between. But by the end of the expedition's first year at sea they had crossed the Atlantic four times, three times north of the equator and once to the south, and had accumulated a mass of data about the bottom sediments. A coherent picture was beginning to emerge.

First, the question of where *Globigerina* and other similar organisms lived was settled, largely through the work of John Murray. Whenever he could, Murray used tow nets (simple, fine-mesh nets that were trailed through shallow water) to collect organisms from surface and near-surface waters. Usually they came back full of plankton—foraminifera and other tiny floating organisms. These were the same creatures that were found as empty shells in the bottom ooze, except that the ones at the surface were alive. Many of them had decorative features such as long, delicate spines that were generally missing from the shells dredged from the seafloor. Murray also discovered that at any particular location there was a near one-to-one correlation between the type of plankton he found in his tow nets and the shells in the bottom ooze. If *Globigerina* dominated in the surface waters, it also dominated in the sediments; if the proportions of various foraminifera were different at another station, the bottom sediments would mirror those proportions. Murray soon came to the inevitable conclusion that the deposits on the seafloor must be made up almost exclusively of the shells of organisms that lived near the sea surface. When the organisms died, their remains

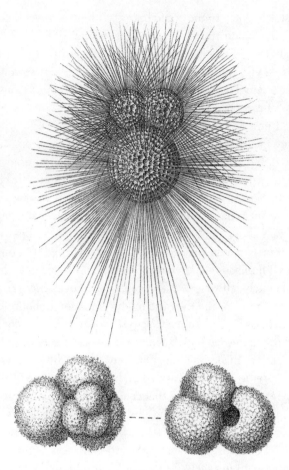

Magnified views of *Globigerina bulloides*, generally the most abundant foram species in *Globigerina* ooze. The upper drawing by J. J. Wild is of a specimen collected during the expedition from a surface tow net. It was alive when captured and still had its delicate spines. The lower drawing shows two views of a shell separated from a sample of *Globigerina* ooze; its spines dissolved as it sank to the seafloor. *Globigerina bulloides* shells (without the spines) are typically a few tenths of a millimeter—about one hundredth of an inch—across. (Drawings courtesy of the Centre for Research Collections, University of Edinburgh.)

drifted slowly down to the ocean bottom. This had never before been conclusively demonstrated. Many scientists, including Thomson, had previously thought that most of the ooze-forming organisms must live at the sea bottom. But Murray's observations were so compelling that once again Thomson had to change his thinking. "Considering the mass of evidence which has been accumulated by Mr. Murray," he said, "I now admit that I was in error."

Murray's work solved the question of how the calcareous deposits originated, but there was still the issue of where the fine-grained, reddish-brown mud came from. After *Challenger*'s multiple crossings of the Atlantic, the dredging records showed that it was a widespread and important type of deep sea sediment. Wherever the depth was greater than about 15,000 feet—roughly 2.8 miles—there were very few foraminifera, and the red-brown mud prevailed. On one of the crossings Thomson tabulated its presence or absence in each dredge haul and concluded that *Challenger* had sailed over *Globigerina* ooze for about 720 miles, and over the reddish mud for 1,900 miles.

But what was it? Unlike the *Globigerina* ooze, it had no connection with living organisms. John Buchanan analyzed the mud and discovered that it was made up almost entirely of very fine grains of clay. Its characteristic reddish-brown color, he found, was due to abundant oxidized iron and manganese that coated the grains. Basically it was a kind of rusty clay, and ever since its discovery it has been known as red clay. The *Challenger* scientists surmised that either little organic life existed in the surface water in the red-clay areas—and thus there would be few sinking shells—or, more likely, that in the deep regions where they found red clay the shells of the surface-dwelling organisms had dissolved on their long journey to the seafloor. As a test, Thomson asked Buchanan to treat a sample of *Globigerina* ooze with acid, which would dissolve the calcareous shells. When he did so, all that remained was a small amount of residue that looked just like the red clay. The experiment seemed conclusive, but it still did not answer the question of origin. Was the clay present within the shells, or did it come from somewhere else entirely? Buchanan used the same procedure to dissolve samples of plankton that Murray had collected from near the surface in a tow net. This time no red clay showed up in the undissolved residue. One possibility could be ruled out: the red clay was not integral to the shells themselves.

Eventually Murray came up with an idea: the red clay, he concluded, must be formed by alteration of volcanic material. Nearly everywhere they traveled, but especially later in the voyage when the ship was in the Pacific Ocean, he found small amounts of volcanic debris in the bottom sediments. Often it was present as small crystals; sometimes there were also larger pieces of pumice. It was well known that, on land, weathering of ordinary rocks produces clay. The same thing would happen to any volcanic material in the oceans. Murray deduced that the red clay must have formed by the breakdown of pumice.

We now know that much of the red clay originates as fine, wind-blown dust and volcanic ash, not discrete pieces of pumice, but Murray was basically correct: red clay is mostly altered volcanic material. It is present almost everywhere, but over much of the seafloor it is overwhelmed by biological debris. Red clay dominates only in the deepest parts of the ocean, where sinking biological material dissolves before it reaches the seafloor. It accumulates extremely slowly, typically at a rate of a tenth of an inch or less every thousand years, which is why red clay areas are such fertile hunting grounds for cosmic dust: proportionately the extraterrestrial particles are much more abundant in these places than elsewhere.

The discovery of extensive areas of red clay in the Atlantic meant that the seafloor could no longer be thought of as one vast depository of chalk deposits. But nevertheless there *were* large regions where calcareous sediments blanketed the ocean bottom; the same processes that had produced Cretaceous formations like the White Cliffs of Dover were still going on, essentially unchanged, in many parts of the ocean. The calcareous ooze and the Cretaceous chalk had similar physical and chemical characteristics, and they also had close biological ties.

The various species of *Globigerina* found in present-day ocean sediment are different from those in the Cretaceous chalk, but they are closely linked biologically: they belong to the same genus—*Globigerina*. When Thomson was a young man learning natural history, the consensus among biologists was that species do not change. How they originated was a bit of a mystery; they were "created"—by God according to some, although not everyone was convinced of this—at a particular place and time when conditions were right. Every subsequent individual of a species was thought to have descended from that original parent or pair of parents. If a species was widespread, well, it had migrated and expanded from its original place of creation until it ran into a natural barrier such as a mountain range or encountered some other inhospitable condition. If that were the case, the similarity between species in the Cretaceous chalk and the Atlantic *Globigerina* ooze indicated only that they had been created to occupy similar ecological niches, where the environmental conditions were the same or nearly so. But by the time *Challenger* sailed, these ideas had been challenged. Some fifteen years earlier Charles Darwin and Alfred Russel Wallace had published their theories about evolution through natural selection, upending long-held beliefs about the natural world. Thomson wrote in 1873 that there had been a "very great change of opinion within the last ten or twelve years" as a result of Darwin's and Wallace's work. It was no longer necessary to think about different species being independently created; they could be linked through evolution. Thomson described it as "descent by modification": over very long periods of time a species could be nudged in one direction or another through small changes in its environment. If these nudges persisted, the cumulative changes would result in a new species and the old one would become extinct. In this scenario it would be possible to trace a continuous line of descent between an extinct species in the Cretaceous chalk and an extant one in the Atlantic ooze.

If all species originated by descent from preexisting organisms, then somewhere back in the mists of time there must have been a single original ancestor. The fossil record showed that as a general tendency, organisms became more complex as time went on. A few primitive organisms existed in the present-day world, such as bacteria and amoebas, but many others occurred only as fossils. A common hope when *Challenger* began her voyage was that some of these apparently extinct organisms still lived in the dark, cold depths of the ocean and would be dredged up during the expedition.

One of the primitive organisms the *Challenger* naturalists were keen to collect and study was a strange creature known as *Bathybius haeckli*. At the time of the expedition, *Bathybius* was the poster organism for the new ideas about evolution by natural selection. It had been discovered in 1868 by Thomas Huxley—the same man who interviewed the young Rudolf von Willemoes-Suhm and recommended him for the *Challenger* expedition. Huxley discovered *Bathybius* in some of the first samples of deep sea mud ever recovered. They had been collected in the North Atlantic by ships surveying the region in preparation for laying undersea telegraph cables, and they had been sent to Huxley for analysis. He found that each sample contained an amorphous, gelatinous substance forming an irregular layer in and on top of the mud. When he examined this strange material through a microscope he noticed that it seemed to incorporate various components of the sediments, including tiny platelets made of calcium carbonate called coccoliths, which are common both in calcareous ooze from the seafloor and in chalk formations like the White Cliffs of Dover. Huxley himself had been the first to observe these minuscule platelets, and had given them their name, but neither he nor anyone else had figured out where they came from or what they were. Now, seeing them embedded in *Bathybius*, he thought they might be an integral part of this new organism, or perhaps food it had ingested. As he observed *Bathybius* under his microscope, Huxley thought he could sometimes see movement. He could not detect a nucleus or other internal organs, but, crucially, when he tried a test that biologists of that time routinely used to determine whether a substance was organic—application of the red dye carmine—the result was positive. To Huxley, this implied that the gelatinous material was organic, and alive. He was convinced he had discovered a primitive organism, and he quickly published his observations, naming it *Bathybius haecklii* in honor of a German colleague, the naturalist Ernst Haeckel.

The news of Huxley's discovery spread like wildfire through the scientific community. Soon a few others with access to deep sea muds reported that they too had observed *Bathybius.* Thomson was one of them, and with his flair for describing things in easily understood terms he wrote that under the microscope *Bathybius* resembled the white of an egg. He also agreed that it was capable of some movement and that it exhibited all the signs of being a form of primitive life. Still, he harbored doubts.

Not so Ernst Haeckel. He enthusiastically embraced *Bathybius* as probably the most ancient and primitive living matter yet discovered. Like Huxley, he was a prominent supporter of Darwin's ideas about evolution, and *Bathybius,* he realized, could be the key to the origin of life on earth, a starting point for evolution. He promoted the theory that the organism might be the link between simple chemicals and true organic life. As more scientists reported finding *Bathybius* in mud from the deep sea, it became the accepted—but nevertheless astonishing—mantra that a great sheet of primeval, living protein was probably spread across the floor of the world's oceans. If Haeckel was right, *Bathybius* had spontaneously become living matter from nonliving precursors, and might have been doing so continuously over much of the earth's history, up to the present moment. It was a profound concept. Still, even Haeckel recognized that much more work had to be done to confirm his theory. He wrote: "We stand here face to face with a series of dark enigmas, the answer to which we must hope to receive from future investigations." Scientists have been wrestling with the question of how life arose on earth—how ordinary chemicals were transformed into organized, reproducing, living organisms—since Haeckel's day. Ideas have ranged from the catalytic properties of clay mineral surfaces to the effects of lightning in an early atmosphere and to the MIT physicist Jeremy England's theories about the possibility that chemical self-organization is inevitable under certain thermodynamic conditions, leading to living

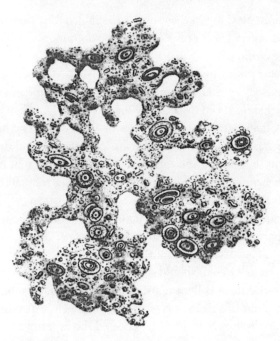

A microscopic view of *Bathybius haecklii* as shown in Wyville Thomson's *The Depths of the Sea*. Thomson credits the image to Haeckel. The amoeba-like nature of this "organism" is obvious. The round or oval objects scattered throughout are tiny coccolith platelets, ranging from a few to about twenty micrometers across (one micrometer is 0.000039 inches).

structures. But Haeckel's enigma has not yet been solved, and his appeal to future investigations still stands

It should not be surprising, then, that investigating deep sea mud for the presence of *Bathybius* was high on the agenda of the *Challenger* scientists. Every sounding and dredge haul that brought up bottom sediment was examined carefully. Murray was especially diligent. He would immediately skim off a thin, watery layer from the surface of each sample of seafloor mud that came on board, and he or one of the other naturalists would observe it under the microscope, often for hours at a time. Two and a half years into the expedition, though, after examining hundreds of samples, they had found no sign whatsoever of *Bathybius*. Nothing in the samples resembled the organism physically, and none of them responded to the carmine stain. The scientists were baffled.

But a clue to the mystery came one day when Murray noticed that there *was* a jellylike layer lying over the mud in some of the specimen jars that had been stored away for future study. When he examined this material under the microscope it looked very much like Huxley's *Bathybius*. Neither he nor his colleagues could detect any movement, but it took the carmine stain, indicating that it was organic. Perhaps they had found the primitive organism after all.

John Buchanan was as interested as the biologists in the questions surrounding the existence of *Bathybius*, but as a chemist he took a different approach. Instead of using a microscope, he performed experiments. He reasoned that if the organism existed everywhere on the seafloor, traces of it should show up in the bottom-water samples that were routinely collected at each observing station. So he took water from one of these samples and evaporated it, expecting to find an organic residue. Nothing. He repeated the experiment on multiple samples, and each time the result was negative. He could find no organic residue that might be the remains of *Bathybius*. The puzzle deepened again. The only possibility, Buchanan decided, was that everyone had it wrong and the gelatinous layer was not a living organism but something completely different.

To test this idea he took some of the gelatinous material from Murray's sample jars and analyzed it chemically. To his—and everyone else's—surprise, his analysis showed that it did not appear to contain any organic compounds at all; it consisted mainly of calcium and sulfate. Then Buchanan tried another experiment: he dissolved some of the jellylike substance in water and evaporated it. As it dried, crystals of calcium sulfate—the mineral gypsum—began to precipitate. Buchanan knew that the same thing would happen if he evaporated ordinary seawater—crystals of gypsum, along with various other salts, would precipitate. Although we usually think of the salt in seawater as being sodium chloride, many other dissolved components are present as well, including large amounts of calcium and sulfate.

Most of the world's gypsum deposits were formed when natural evaporation of seawater in lagoons or shallow seas caused the mineral to precipitate.

Buchanan's chemical detective work showed that *Bathybius* was not a living thing at all; it was simply a gelatinous form of calcium sulfate. But why was it gelatinous? Further detective work revealed that the so-called *Bathybius* was only present in sample jars to which alcohol had been added as a preservative. This was a common practice; the mud samples examined by Huxley—the ones in which he discovered *Bathybius*—had also been stored in alcohol. A little seawater was always present in the mud samples, and Buchanan realized that adding alcohol to seawater would cause the dissolved calcium sulfate to precipitate out as a gel. To convince himself, and the others, he did the experiment: he mixed ordinary seawater with alcohol. Soon an amorphous, jellylike precipitate that looked just like *Bathybius* appeared. The puzzle, it seemed, had finally been solved. The *Challenger* naturalists had not been able to find *Bathybius* in their fresh mud samples because there was no such organism. *Bathybius* was in reality an inorganic precipitate that formed when samples containing residual seawater were preserved in alcohol.

However, one nagging question remained. Why did the gelatinous material always give a positive result in the standard test for living tissue, staining with carmine dye? This had been one of the strongest arguments for believing *Bathybius* was a living organism. Buchanan had an answer for this too. He pointed out that a common industrial process for making pigments was to stain an inorganic precipitate—such as the gel that everyone thought was *Bathybius*—with an organic dye like carmine. The same thing had happened with the so-called *Bathybius*. Yes, carmine dye stained organic tissue, but it could also stain the inorganic calcium sulfate precipitate.

When it became clear to everyone on *Challenger* that *Bathybius* was not a living organism, Thomson wrote to Huxley and broke the news. It was unwelcome, but Huxley, recognizing his mistake, took it in his stride and published a retraction, crediting the findings of the *Challenger* naturalists. *Bathybius* was a cautionary tale. Huxley had been certain of his identification of a new, primitive organism from the seafloor. But he was also a vocal proponent of Darwin's ideas, and *Bathybius* fit nicely into the evolutionary scheme. That, perhaps, had colored his thinking and made him less critical in his research than he otherwise might have been. Still, science progresses through the correction of such errors, and it was the curiosity of the *Challenger* scientists about *Bathybius*, and their determination to understand why they could not find it in their samples, that eventually led them to the truth about the "primitive organism" that wasn't.

Ernst Haeckel, for his part, was not so quick to give up on the idea that *Bathybius* was real, perhaps in part because it had been named after him. For a while he continued to write about *Bathybius* in his publications. He claimed that its distribution might be more limited than originally thought and that perhaps *Challenger* had simply not dredged in locations where it was common. But most of the scientific community recognized the validity of Buchanan's findings, and within a few years *Bathybius* had faded into obscurity.

five

THE SAGACITY OF CRABS

Saint Paul's Rocks, as one might guess from the name, are nothing but a small cluster of rocks. What is unusual about them is that this small cluster of rocks sits in the middle of the vast Atlantic Ocean far from any other land. Their highest point is only fifty-nine feet above sea level and their above-water area is so small that they would fit into an average-size city park. They are too small to be considered islands so they are often referred to as islets, although if you were to remove the ocean you would see that they are actually the peak of a high mountain, its base several miles below the surface. In spite of their small size, though, like every piece of solid ground that sticks up above the ocean's surface, they have attracted sailors and explorers for hundreds of years. Darwin visited them in H.M.S. *Beagle.* The first known human encounter with Saint Paul's Rocks came one dark night in 1511 when a Portuguese ship on its way to India crashed into them. Fortunately the wrecked ship was part of a small convoy; none of the other boats foundered, and the sailors were rescued by their comrades. They must have wondered how they managed to hit such tiny specks of land in the midst of a huge expanse of ocean. But by 1569 sailors were being forewarned; the Rocks appeared on Mercator's nautical map, published in that year.

Challenger arrived at Saint Paul's Rocks at the end of August 1873, eight months into her voyage. She was already on her third crossing of the Atlantic. After leaving Tenerife she had sailed west to the Caribbean, discovering the red-clay deposits along the way, and then north to Halifax, Nova Scotia. From Halifax she crossed the Atlantic once more, eastward to the Azores, and then turned south to Madeira and Cape Verde. Eventually, as she sailed west again toward Brazil, she arrived at Saint Paul's Rocks. She approached them warily, under steam, with the crew making frequent soundings and the captain on high alert for shoals that could tear a hole in the hull. But the depth

Challenger tied up to Saint Paul's Rocks, from a drawing by J. J. Wild. Wild put himself (or another artist) in the picture at lower left, and nearby he even sketched in a few of the ubiquitous crabs, *Grapsus grapsus*, that were among the most notable inhabitants of the Rocks. (Courtesy of the Centre for Research Collections, University of Edinburgh.)

drops off so quickly around the Rocks that the ship was able to maneuver in close enough to tie fast to a protruding rock, sheltered from the prevailing wind and current. She stayed there for two days and was, to say the least, an unusual sight. This was almost certainly the first time a large ship had been moored to a small group of rocks in the middle of the Atlantic Ocean.

Saint Paul's Rocks could not be considered a port call for *Challenger*, but it was land, however small, and the captain gave the crew a day's holiday. All who wished could go ashore to stretch their legs, and most did—it is likely that on August 28, 1873, there were more people on Saint Paul's Rocks than have ever been there before or since. Fishing was a great attraction (recall Willemoes-Suhm's amusement at his shipmates' enthusiasm for this activity), and in a short time the fishermen—sailors, officers, and scientists alike—caught enough fish to feed everyone aboard (during the two-day visit they also managed to add seven new fish species to *Challenger*'s collections). Later, as was the custom when visiting remote places, they left a message for future travelers. It listed the officers of *Challenger* and gave details of the magnetic measurements they had made. Then it got to the important part: "Caught plenty of fish." Henry Moseley was one of the naturalists who liked to fish; he had taken care to pack his fly rod and other tackle for the voyage. But the Rocks were the first remote place the naturalists had visited. Their main concern was to document and collect every living thing they could find.

At the time of *Challenger*'s visit, Saint Paul's Rocks were disputed territory. Today the islets are officially part of Brazil, and they have been renamed, rather grandly considering their size, Saint Peter and Saint Paul's Archipelago—SPSPA for short. In 1986, because of the uniqueness of the archipelago, the Brazilians declared it an environmentally protected area. Tourism is not permitted, but a permanent research station was established in 1998, and since then the Rocks have been inhabited by scientists working on the rich ecology, geology, and oceanography of the region. Why are these tiny islets special enough to merit establishment of a permanent research station? Cynics might point out that building it allowed Brazil to add a two-hundred-mile economic exclusion zone around the archipelago. But the scientific justifications for the research station are compelling.

The *Challenger* naturalists soon recognized one of the features that make the islets unique: unlike almost every other known oceanic island, Saint Paul's Rocks did not appear to be volcanic. Darwin, forty years earlier, had come to the same conclusion. But neither Darwin nor the *Challenger* scientists could determine how they were formed. It was not until a century later that the origin of the small islets began to be understood in the context of plate tectonics.

None of the *Challenger* naturalists was a trained geologist, but they nevertheless carried out the first thorough geological survey of Saint Paul's Rocks. They also collected representative rock samples that were later examined comprehensively by geological colleagues. From their on-site study, through both visual observations and John

Buchanan's chemical analyses, they concluded that the islets were predominantly composed of a rock they called serpentine. (Strictly speaking, serpentine is a mineral, not a rock, but rocks that are rich in the mineral serpentine are often themselves referred to as serpentine. For simplicity I'll follow that practice here.) Serpentine is not an especially common rock, but because of its beguiling greenish colors and beautiful polished surfaces it is a favorite of sculptors and jewelry makers, and is sometimes used as a decorative building stone. (It is also the state rock of California.)

Serpentine was well known to geologists of the 1870s as an interesting rock type of uncertain origin. After examining the serpentine at Saint Paul's Rocks carefully, the *Challenger* scientists agreed with Darwin's earlier conclusion: the rocks did not have a volcanic origin; they had been formed some other way. They were correct: we now know that serpentine is metamorphic, not igneous, and originates in the earth's mantle, deep below the surface. Serpentine found at the earth's surface, like that at Saint Paul's Rocks, has been thrust up from deep below by processes related to plate tectonics; in the case of the Rocks, a huge block of mantle material was raised up along a fault zone—incrementally and very slowly from a human perspective—until the islets, the jagged peaks on the top of the block, reached their present position several miles above the seafloor. We know these details because since the *Challenger* expedition, and especially in recent decades, the region around the Rocks has been surveyed intensively, dredged repeatedly, and even examined close up from a manned submersible.

Saint Paul's Rocks sit on a fault zone that cuts across the Mid-Atlantic Ridge, the relatively shallow feature first delineated in detail by the scientists on *Challenger*. The Ridge is a long, sinuous, elevated region that snakes through the length of the Atlantic Ocean, dividing it neatly in half; it is a scar on the ocean floor, tracing the line along which continents on either side of the ocean split apart more than a hundred million years ago. (The existence of shallow regions in the central Atlantic had been known before the *Challenger* expedition, but the ship's multiple crossings and regular soundings showed that it was a continuous feature.) The bits of the mantle that we now know as Saint Paul's Rocks may have been torn out of the earth's interior during the continental breakup that initiated the formation of the Atlantic Ocean; their journey up to sea level happened much later by periodic movement along the fault, which is still seismically active today: a strong earthquake hit the region in 2006 and did considerable damage to the research station on the Rocks.

The *Challenger* naturalists knew none of this. Their description of the islets as being made of serpentine was a label, not a term with implications for their origin. The scientists were curious about the geology mainly because they knew it was unique among oceanic islands, but as interesting as that geology was, they were even more curious about the zoology and botany. They had all read Darwin's report from his earlier visit. Now they were ready to do their own investigation, to confirm what he had discovered and if possible add to his list of species.

The Rocks sit close to the equator and are constantly buffeted by the southeast trade winds blowing from Africa. At the same time the South Equatorial Current, also coming from the east, sweeps by them at up to several miles per hour. The *Challenger* scientists, and almost everyone else on the ship, were struck by the power of the current as it pelted by the Rocks "like a mill run." It flowed like a river, and it made handling the ship's small boats difficult whenever they ventured beyond the sheltered lee side of the Rocks. What the scientists were unaware of was the presence of a countercurrent flowing in the opposite direction, west to east, around one hundred feet below the surface. The combination of the trade winds and ocean currents flowing both east and west is one of the features that make Saint Paul's Rocks unique: organisms can be carried toward the Rocks from both Africa and South America, and can use the Rocks as a way station in mid-ocean, facilitating biological exchange between the two continents. Like any land poking above the sea surface, the Rocks also provide temporary and occasionally permanent refuge for migrating creatures or, indeed, any organisms that arrive there—especially seabirds. And the submerged slopes of the islets are a mid-ocean haven for fish and other small creatures that live in shallow water. But in spite of these favorable conditions, Saint Paul's Rocks—like many small, remote islands—lack biological diversity. The *Challenger* scientists found only a few varieties of algae, some ticks, three species of bird, various fish and other marine creatures, and numerous crabs. By the standards of other tropical islands they visited, their collection of living things from the Rocks was meager.

Among *Challenger*'s naturalists, Moseley was the most intensely interested in the natural history of islands. At sea he was as busy as the others examining and describing creatures brought up in the trawls and dredges, but when the ship arrived at an island he was in his element. He was in the first boat to go ashore at Saint Paul's Rocks, as he was at almost every island visited during the expedition. The islets form a roughly semi-circular bay on their western side, protected from the prevailing easterly winds and currents. No suitable anchorage existed, but in this relatively calm bay it was possible to tie the ship to the Rocks with ropes. Even in the bay, though, the large Atlantic swells swept in and out, rising and falling impressively against the steep rocks of the shoreline and sending spray everywhere. So landing on the jagged rock outcroppings was tricky. When Moseley went ashore he had watch carefully and time his jump onto the rocks to the precise moment his boat was at the peak of a wave, then "cling to the rocks as best [he could]." His attire probably did not help. He and the other civilians on the ship usually dressed in shirts, ties, vests, and jackets, both at sea and during field excursions on land. Their outfits, in fact, were not vastly different from their dress back at the University of Edinburgh or Oxford, although in the tropical heat at the Rocks they may have been induced to shed a few layers. But the first landing party made it ashore without incident and once there were able to help the groups that followed to an easier landing, although Wyville Thomson commented that it still had to be accomplished "with a spring and a scramble when the boat is on the top of a wave."

The expedition had no official botanist, so Moseley adopted that role and was the primary collector of land plants throughout the voyage. In fact, he seemed ready to take on any task. He had his own interests, of course, but he once commented that he would happily fill in any investigative gaps left by the other scientists. As botanist on the inhospitable Saint Paul's Rocks, however, he had little luck. There was no vegetation at all, "not even a lichen," he lamented in his journal. On his search for things green, he did manage to find a species of microscopic algae. In sheltered places it grew on the guano, the white bird dung that coated most of the rocks. He found other varieties of green algae in the stagnant pools that lay among the rocks, and carefully sampled and documented each species. Many of these samples—including the microscopic algae—still form part of the *Challenger* collections at the Natural History Museum in London.

What the islets lacked in vegetation they made up for in birds. As the ship approached, Moseley noted that birds could be seen "in thousands" flying over and around the Rocks. Like the scientists, the birds were curious. They were soon circling *Challenger*, giving it a thorough examination. As soon as he had clambered onto the shore, though, Moseley in turn began examining the birds: their nests, their eggs, their chicks, and the adult birds. There were only two varieties, noddies and boobies, English names that sound as if they came straight from a children's storybook. As he worked they constantly circled overhead, "screaming in disgust" at his invasion of their territory. But on the ground and especially on their nests they were docile, and from all appearances not particularly adept at self-protection. Except for the occasional visiting sailor, there were no predators on the Rocks. The birds made no protest when the *Challenger* naturalists approached, and they could be knocked over with a stick. Moseley says he could have caught as many noddies as he wanted with his bare hands, and the boobies were just as unsuspecting. When he walked up to a nest to examine it, the birds simply sat there; the only way he could find out how many eggs they had was to push them off. Mosely was not the first to find the behavior of these birds obtuse; the name "boobies" probably comes from the Spanish slang *bobo*, "stupid." The noddies, too, have long been viewed as not very bright; their Latin genus name (*Anous*) is derived from an ancient Greek word that also means "stupid" (their English name probably comes from their habit of bobbing up and down to one another during courtship).

Neither noddies nor boobies were rare or unusual, but the naturalists nevertheless collected many specimens for further study. It was near the end of the breeding season, so they were able to find all stages of development: adults, chicks, and eggs, for their collection. One of their goals was to examine the specimens for varieties that were unknown elsewhere; the Rocks were so isolated—the nearest land, itself another small archipelago, is four hundred miles away—that this was a distinct possibility. The naturalists also observed and recorded details of the birds' behavior. Moseley, for example, writes about the precarious looking nests, built up on rock ledges with layer upon layer of seaweed and feathers cemented together with dung. He judged that they had been used by generations of birds, each new inhabitant adding another layer to the nest. But he also notes that the high bird population resulted in fierce competition for space, and many females had to lay their eggs on bare rock. Whenever possible they chose a small hollow so the eggs would not be blown away by a gust of the unceasing wind.

The birds' primary food was fish, and any unoccupied space surrounding their nests was littered with fish debris. Sometimes the birds regurgitated an odoriferous porridge of ground-up fish for their young, not all of which made its way into the expectant mouths. Fresh excrement coated the rocks everywhere, and the stench was overpowering. But nothing deterred the *Challenger* naturalists. With birds dominating the biology of this tiny, isolated piece of land, they knew that most other living creatures on the Rocks would have to be connected with them. So they scraped and scratched in the guano, teased apart nests, scrounged through the feathers and fish skeletons, and probed the carcasses of the occasional dead bird. It was not very pleasant, and not very productive. They came up with a couple of species of spider, a tick that lived on the boobies, and a tiny creature that looked like a miniature scorpion but was actually in the same class of animals as spiders—a pseudoscorpion. They also found a parasitic fly and lots of lice. None of these may sound exciting, and for some readers conditions in and around the birds' nests might seem disgusting. But the naturalists lived for their collections; almost nothing could distract them from their goal. They enthusiastically accumulated multiple specimens of everything they could find, a practice they continued throughout the voyage. They shot birds and stuffed them, scraped algae from rocks, collected spiders and lice and preserved them in alcohol, and caught fish.

In the meantime several of the naval officers were busy surveying. Their aim was to assess whether it would be feasible to build a lighthouse on the Rocks. They were British, after all, and the British Navy ruled the seas. A lighthouse on these remote islets would warn sailors of danger, helping them avoid the fate of the Portuguese ship that crashed into the Rocks in 1511. It would also serve as a useful marker, a fixed point in the ocean where mariners could adjust their navigational instruments. In spite of the small size of the islets, the surveyors identified a relatively flat area that was big enough to accommodate a lighthouse. Its construction would displace a few birds but that did not concern them. The practicalities, however, were daunting. There were no natural resources. A lighthouse would need to be manned and regularly supplied with fuel, food, and fresh water. What might at first have seemed like a good idea fizzled quickly in the face of reality. It would be another fifty years before a lighthouse was built on the Rocks—by the Brazilians—in 1930.

While others were busy on land, Thomson and Murray tried a novel type of dredging from the stationary ship, still moored fast to the Rocks. They had tested the method earlier, in Bermuda, and it had worked well. It involved sending out the dredge in one of the small boats, paying out the line as the boat moved away from the ship and then dropping the dredge to the bottom. As the winch operator on *Challenger* slowly wound the dredge back in, it scraped over the submerged rocky outcrops on the submerged flanks of the islets and scooped up multiple specimens of coral, sea fans, and crustaceans to add to the expedition's collection. This minor episode is a nice illustration of how the scientists wasted no opportunity to make observations and collect samples.

The most notable denizens of the Rocks were the crabs: *Grapsus grapsus*. Almost everyone who has visited Saint Paul's Rocks has commented on them. Crabs are abundant in and around the sea, particularly in the tropics, and *Grapsus grapsus* is the most common variety along both the east and the west coasts of South and Central America, as well as on many islands. On Saint Paul's Rocks it is the only crab species and, after the noddies and boobies, the crabs are the largest living creatures. Moseley noted that they appeared in "vast abundance" on the Rocks, and they provided endless fascination—and sometimes frustration—for the visitors. "Cheeky, exasperating, but intensely amusing," as George Campbell described them.

Grapsus is also known as the red-rock crab, or sometimes the Sally Lightfoot crab, the latter a reference to its extreme agility. When the *Challenger* scientists landed at Saint Paul's Rocks, the crabs were crawling about everywhere. But when the naturalists tried to catch them, or even to pay close attention to them, then *poof*—in an instant they disappeared into a crevice. They were, however, appreciative of the *Challenger* fishermen's practice of cutting up small fish to use as bait. When the crew first arrived, without thinking twice they would put their bait down on a rock beside them while they fished. But as soon as they turned their backs a hungry crab would snatch it away. The fishermen soon learned to be more attentive. It was much easier to keep a close eye on their bait than to retrieve it from a wily crab. Even large, freshly caught fish, left unattended, were fair game. They would very quickly be swarmed over by crabs plucking, clawing, and tearing away morsels of flesh.

Absent free meals from visiting fishermen, the standard food of *Grapsus grapsus* is algae. If you search for *Grapsus* on YouTube, you can find videos of these crabs deftly plucking bits of algae from the volcanic rocks of the Galápagos Islands (where they are also abundant) and methodically stuffing them into their mouths, left claw, right claw, left claw, repeat. But on Saint Paul's Rocks not even algae is abundant. With a large crab population, competition for food is intense. Most of the time the crabs are omnivorous scavengers, eating algae when they can find it, but also whatever else is available. The abundant birds on the Rocks are a godsend for them; around the nests the crabs can find bits of fish, regurgitated fish meal, and the occasional dead boobie or noddie, which they clean off down to the bone. They even nibble on the ubiquitous bird dung. Moseley watched a large crab grab a recently hatched chick from its nest and scurry off. Reportedly, they also eat the birds' eggs. They are indiscriminate and cannibalistic; large mature crabs will eat smaller ones.

Like the birds, the crabs have few predators on the Rocks, which accounts for their large numbers. Occasionally a boobie will eat a small crab, but it will not tangle with an adult; those pincers are too dangerous. About the only other thing the crabs have to worry about is getting washed into the sea. If this happens they are likely to become lunch for a fish. George Campbell was no friend of the crabs; he had experienced firsthand their skill at stealing his fishing bait. For hours he tried to capture one, and when his persistence finally paid off, he hurled the poor creature into the ocean and watched, satisfied, as a fish swallowed it. Revenge, he declared, was sweet. However, angry seamen aside, it is rare for a crab to be washed into the sea. They have such a powerful grip that even heavy surf cannot dislodge them; when a wave sweeps over them they flatten themselves against the rocks and cling on tenaciously, never loosening their hold. (This is another behavior documented in the Galápagos YouTube videos.)

The adult crabs, especially, are beautiful. Although drab as youngsters—dark browns and greens prevail making them almost indistinguishable from the rocks—they molt frequently as they grow, gradually becoming brighter, until the mature crabs sport brilliant reds, yellows, and blues. Compound eyes, projecting from their head on stalks, have an almost human quality, appearing to fix their gaze intently on the observer. Their vision is acute, and their reactions are instantaneous; one false move and the crab races off in whatever direction will most easily take it to safety. Their popular name Sally

Lightfoot is well earned. When it senses danger a crab will speed across the rocks on its four pairs of legs, leaping across crevices and scurrying up almost vertical rock faces. Campbell swore he watched an old crab vault over a crack at least two feet wide and land easily on the other side. Sometimes they draw themselves up on tiptoe and strut about as if to say, Look at me. Moseley, his curiosity piqued, watched their behavior closely. They had, he said, a peculiar way of expressing emotions. From the attitude of their claws he thought he could detect anger, astonishment, suspicion, and fear. His encounter with the creatures on Saint Paul's Rocks completely changed his opinion about them. Prior to his visit, he admitted, he had not thought crabs very intelligent. He could not say why he held that belief, but after observing *Grapsus grapsus* on the Rocks he had an epiphany. He was, he wrote, "astonished at their sagacity."

six PENGUINS GALORE

On her fourth crossing of the Atlantic, sailing from Brazil to Cape Town, South Africa, *Challenger* arrived at Tristan da Cunha, a volcanic island in the South Atlantic. Unlike Saint Paul's Rocks, Tristan is a large, proper island boasting permanent inhabitants. It also hosts penguin rookeries, and Moseley, who was fascinated by penguins, rushed ashore to find them. But the logistics of getting to the nesting areas proved complicated, and the captain had warned those going ashore that they would have to stay within sight of the ship because if bad weather rolled in—a distinct possibility in that part of the world—he would hoist the recall flag. The anchorage at Tristan da Cunha was treacherous; the ship would have to raise anchor and leave if a storm blew up. Shore parties needed to remain within an hour's walk of the landing place. Disappointed but philosophical, Moseley decided to use his time "botanizing" rather than chasing penguins. Nevertheless he alerted the locals that he was willing to pay if they could bring him a pair of penguins and their eggs. He soon discovered that the islanders were shrewd negotiators. By time he was ready to leave he had his penguins and eggs, but he had paid dearly for them. Even so, in his journal Moseley urged others who might find themselves in similar circumstances to follow his example: "Had I known what countless numbers [of penguins] I was so soon to be amongst I should not have made such an offer, but I have found in the long run that, on a voyage like this, where there is so much uncertainty, it is always best to take the very first opportunity, and I always landed on the places we visited with the very first boat, even if it were only for an hour in the evening. It may come on to blow, and another chance may never occur. I strongly advise any naturalist similarly situated to do the same."

When *Challenger* anchored at Tristan da Cunha, Campbell commented, "The appearance of the place makes one shudder." The soil

93

was poor, the climate uninviting, and the entire population numbered eighty-six persons. The main—and only—settlement was called Edinburgh of the Seven Seas, in honor of a visit by Queen Victoria's son Albert, duke of Edinburgh, a few years earlier. The village, a handful of houses cowering under the massive black cliffs of a snow-topped volcano, was a far cry from the Edinburgh many of those on *Challenger* knew. The residents were a mixed lot, descendants of whalers and shipwrecked sailors, and a few who counted among their ancestors members of the original military unit that had been sent to the island in 1816, when Britain annexed it—ostensibly to ensure that it was not used as a staging point for an attempt to free Napoleon from his exile on Saint Helena, some fifteen hundred miles to the north. Today Tristan is a British overseas territory, with a resident population somewhere between 250 and 300 people. It has no airport and can be accessed only by ship, typically via a six- or seven-day journey from Cape Town. It bills itself as the world's most remote inhabited island. Its main agricultural product, both at the time of the *Challenger* visit and now, is potatoes. In 2016 the National Farmers Union in the United Kingdom advertised for an agricultural advisor, a hands-on farmer, to help the islanders expand their agricultural horizons. An attractive package was available, the advertisement promised, including free travel and accommodation. The duration of the assignment would be about two years—but its exact length would depend on the shipping schedule.

Moseley crammed as many activities as he could into the short time he had on Tristan da Cunha, and in his journal he describes the geology of the cliffs around the settlement and notes the temperature of the water in local streams and ponds. But he concentrated on the island's biology. The flora and fauna were already reasonably well known, but—being Moseley—he wanted to verify the existing data for himself and if possible add to the list of known species. He spent the morning on the lowlands at the base of the mountain cliffs, examining and collecting plants, grasses, mosses, ferns, and any other vegetation he could find. Later, with a young local boy as his guide, he began to make his way up the slope behind the settlement, hoping to investigate the flora present at higher elevations, which he assumed might differ from the plants on the lowlands. But as they climbed a sudden squall came up, black clouds scudding in from the sea. It began to hail, chilling them to the bone, and to Moseley's amazement the boy flung himself to the ground, burrowed under the tall grass and ferns, and curled up in a fetal position, knees drawn up and head tucked in. After a moment's hesitation Moseley did the same, and was astonished at the degree of protection the "scanty herbage" provided from the storm. It was a lesson in adopting the practices of local inhabitants. When the hail stopped he had to hurry back to the ship—the recall flag had been raised—pausing only long enough to collect one more specimen, a tree fern.

As soon as everyone was back on board, the ship steamed away. While Moseley was on shore the crew had loaded up with fresh beef, lamb, fowl, and—of course—potatoes purchased from the islanders, and now the ship headed for nearby Inaccessible Island, just twenty miles away, the second of the four islands that make up the Tristan da Cunha group (the others are tiny Nightingale Island, which is also very close to Tristan, and Gough Island, some 200 miles to the south). Like Tristan, Inaccessible is volcanic in origin, its steep dark cliffs rising sharply from the sea, and when *Challenger* arrived she anchored off a small beach for the night. Finally, Moseley would get to see his penguins. All night he could hear them "screaming" on shore. Groups of the birds flashed through the water near the ship, leaving glowing phosphorescent trails behind them.

Inaccessible had no permanent settlement, but the people on Tristan da Cunha told the *Challenger* crew that two German brothers were currently living there. They had been put ashore from a passing whaler two years earlier hoping to make their fortune harvesting seals. This was not particularly welcome news to the people of Tristan, because they considered Inaccessible part of *their* territory, and they regularly hunted seals there themselves. They had had no news of the brothers for many months, but when morning broke the crew of the *Challenger* soon spotted the two Germans, standing on the beach, gazing forlornly seaward. A small boat went ashore to

bring them to the ship; they were in reasonable health, but their seal-
ing venture, they said, had been a dismal failure. Theirs was a story
of two years of intermittent hardship with little to show for it, and
they asked to be taken to Cape Town, the next port call for *Challenger*. In the meantime they were a boon for the naturalists: the men
spoke English well and knew the island intimately. Soon Moseley,
true to form, was heading for shore with one of the brothers as his
guide. As their small boat drew close to the island Mosely noticed a
school of what he at first thought was a species of pygmy dolphin leaping in and out of the waves. He estimated there were at least fifty
of them, with black backs and white tummies, heading directly
toward the shore. But the "dolphins" did not stop when they reached
land; they swam right through the surf and clambered up onto the
stony beach. Only then, to his delight, did Moseley realize they
were penguins.

Moseley had read extensively about penguins, but he had had no
experience with them in the wild. He had not realized how at home
they were in the sea. He later wrote that he would never have believed
they were birds had he not seen them climb onto the land. Elegant
and acrobatic as their movements were through the waves, their lo-
comotion on shore was awkward. With both feet together, they
hopped from boulder to boulder, and they continued to hop even
when they reached flatter ground. Moseley's impression was that they
looked like a group of men in a sack race. Small wonder their common name is "rockhopper" penguins. Close up they are beautiful,
mostly dark gray and white, with small splashes of contrasting color:
bright red eyes and beak, and long yellow plumes extending out from
either side of their head, framing their faces and resembling exotic,
oversized eyebrows or an elaborate coiffure.

The Tristan da Cunha group of islands is home to only one variety
of penguin. Formally known as the Northern Rockhopper Penguin,
its scientific name is *Eudyptes moseleyi*, after Henry Moseley. When
Challenger arrived at Inaccessible Island in mid-October, 1873, Moseley
estimated that there were "millions" of penguins in the rookeries.
He may have been exaggerating, but possibly not by much. Toward

the end of the twentieth century, however, the population began to decline drastically, probably due to overfishing and environmental degradation, and today Northern Rockhoppers are classified as an endangered species. Most of the world's remaining population still nests on the islands of Tristan da Cunha, and although in the past Tristan islanders regularly feasted on penguins and their eggs, they now fiercely protect them. In March 2011 a cargo ship ran aground on the rocks of Nightingale Island, and as the weather worsened she broke up before salvage teams could reach the site. Fortunately all the sailors were rescued, but fuel oil leaked out into the sea, creating a huge slick around the entire island; some oil also reached neighboring Inaccessible Island. The feathers of any penguins venturing into the sea were quickly covered in a viscous, sticky layer of petroleum. Volunteers from Tristan da Cunha and a conservation team from Cape Town, bringing specialized supplies and twenty tons of frozen sardines to feed the stricken rockhoppers, worked to rescue as many penguins as possible. Thousands of the oil-coated birds were taken from Nightingale Island to Tristan for cleanup—no mean feat in small boats under difficult weather conditions—after which they were released back into the sea. Pictures of the helpless birds hunched over like little old men, matted with black oil, no sign of their colorful plumage visible, are heart-wrenching. Virtually the entire human population of Tristan da Cunha took part in the rescue operation, and the following year the Royal Society for the Protection of Birds, in a first, gave its prestigious medal not to an individual but to the community of Tristan da Cunha islanders.

When Moseley went ashore on Inaccessible Island he got his first closeup look at the rockhoppers that would eventually be named after him. For their rookeries, the penguins favored parts of the island covered in a tall, dense variety of grass that grows in clumps, known as tussock grass. It provided them with degree of protection from the weather and predators, and in turn their waste was a constant source of fertilizer for the grass, which consequently grew "higher than a man's head." Moseley imagined the rookery as a small city or subdivision. When the penguins arrived from the sea, he said, they mostly landed in one spot and followed a "smoothly beaten black roadway" that led to the "main street." Other streets went off laterally on both sides of this main road. Because the grass was so high, as soon as he entered the rookery he lost all sense of direction. It was, he said, like a maze. Although the German guide with him knew every street and intersection, it was a nerve-racking experience: "It is impossible to conceive the discomfort of making one's way through a big rookery. . . . You plunge into one of the lanes in the tall grass which at once shuts out the surroundings from your view. You tread on a slimy black damp soil composed of the bird's dung. The stench is overpowering, the yelling of the birds perfectly terrifying. . . . You lose the path . . . you are, the instant you leave the road, on the actual breeding ground. . . . You cannot help treading on eggs and young birds at almost every step."

At first Moseley tried to avoid crushing the nests and eggs but soon found it impossible. If he wanted to get through the rookery, he had no choice but to "have recourse to brutality." The penguins, though small, were themselves brutal, and he was not prepared for their viciousness. If he stepped close to a nest, the parent penguin would yell raucously at him, its red eyes gleaming and its plumes "quivering with rage." If Moseley's legs were within range the bird would bite. Sometimes several birds would attack simultaneously, and their beaks were sharp. He was almost maddened by the pain, the stench, and the noise; after his first experience he made sure to wear protective gear whenever he ventured into a penguin rookery. The next day *Challenger* sailed to nearby Nightingale Island, where Moseley surveyed the shore, spotting another vast penguin rookery and "a great drove of penguins" on the rocks. Immediately, he wrote, he went below to put on the thickest pair of leather gaiters he could find.

Tristan da Cunha was *Challenger*'s only stop on the crossing from Brazil to Cape Town, and her stay lasted just a few days. The timing—October, the Southern Hemisphere's spring—was fortuitous. Had the visit been at a different season, Moseley might not have encountered

J. J. Wild's sketch of rockhopper penguins (*Eudyptes moseleyi*) among the tall tussock grass on Inaccessible Island. The birds look benign here; another of Wild's sketches shows one of the sailors violently kicking at penguins as he tries to make his way through the rookery. (Courtesy of the Centre for Research Collections, University of Edinburgh.)

his rockhopper penguins at all, and he certainly would not have been able to observe nesting birds with their eggs and young. He learned about the cycle of penguin life from the two brothers who had been living on Inaccessible, Gustav and Frederick Stoltenhoff. They had studied the habits of the penguins in considerable detail, perhaps in part because penguins and their eggs were main staples of their diet and understanding the birds' habits was important for their own survival. Later they wrote down their observations; they were probably among the first people to do so. In effect they conducted a two-year field study, although this was obviously not their primary activity on the island. Still, few modern researchers are able to carry out such an extended study, and the brothers' observations have remained valuable for others studying the birds' life cycle.

Each year, the brothers reported, the penguins vanished from the island near the middle of April. It happened literally overnight: penguins were present when the two Germans went to bed in the evening but gone the next morning—except for "two or three in sick quarters." The birds did not come back for nearly four months, during which time the brothers were forced to rely on other sources of food. The penguins began to return around the last week of July, initially in small numbers, but gradually swelling to a steady stream. For a few weeks only males appeared, well fed and fat from their time at sea, and at first inclined to do little except sleep and laze around. They never once returned to the water. Eventually, however, they began to prepare nests, round and neat, sometimes built up with tussock grass and sometimes simple circular depressions in the ground, scraped out with their claws. The females started to come ashore about the middle of August, and they made straight for the nesting areas, mated, and within a few weeks were laying eggs. Frederick Stoltenhoff said he was struck by the "evident joy" the penguins expressed when they rushed ashore to greet their mates, "flapping each other with their wings and caressing each other in an unmistakable manner."

By September most of the nests contained two eggs, and both parents tended to them. While one was on the nest, the other would be at sea in search of food, often for more than a week at a time. Then they would switch. According to the brothers, this went on for five or six weeks until the eggs hatched in October and November. During the entire time the Germans dined almost solely on penguin eggs. They fried them in fat from the introduced pigs that roamed the island; like the brothers the pigs were voracious consumers of penguin eggs. The mid-October timing of *Challenger*'s arrival at Tristan, coinciding closely with peak hatching time, accounts for Moseley's gruesome description of stomping through the rookery on Inaccessible, squashing both eggs and young underfoot.

With hungry young penguins in the nests there was a sudden increase in the demand for food, and the adult penguins had to redouble their efforts to forage for the crustaceans and small fish that constituted the bulk of their diet. They rarely took to the sea individually; they nearly always hunted in groups like the one Moseley had mistaken for a school of small dolphins. When they returned to the island the penguins would regurgitate their catch into the waiting mouths of their offspring. By December—the Southern Hemisphere's summer—the adults would finally get a break: the chicks were strong enough to make their own way to sea, and for a few weeks virtually the entire population, chicks and parents both, would desert the island again. The young were learning how to hunt for their own food, and both young and old were gorging themselves in preparation for the next stage of their annual cycle: molting.

A molting penguin is a sorry sight, a patchwork quilt of feathers looking a bit like a child's shabby, well-used stuffed animal. The birds have to endure this indignity every year as they replace their old, worn out-coats of feathers. It is a crucial process, worth the indignity; after a few weeks of scruffiness they emerge resplendent in a sleek new tuxedo, warm and waterproof. Penguins are unusual among birds in that over the few weeks of molting they replace every single one of their feathers; most other birds molt gradually, a few feathers at a time. For the penguins, manufacturing new feathers and keeping warm as they lose their insulating coats requires a tremendous amount of energy. Without their feathers they are neither waterproof nor well insulated, so they cannot go out to sea to feed; instead they have to rely on the store of energy built up during their pre-molt feeding frenzy. As a result their weight drops dramatically.

This behavior is hard-wired into the penguins' biology; even among those kept in zoos, with a guaranteed food supply and no need to hunt, the same sequence of events occurs. As the time for molting approaches their appetites become voracious and they snap at their keepers at feeding time, pushing one another out of the way in the rush to get food and build up their reserves of energy. Once molting is under way, their appetites wane. When offered food they eat little; they are quiet and morose, conserving energy, waiting out the time until they are once again fully feathered. Just like their relatives in the wild, the zoo penguins can lose a third or more of their pre-molt weight during this period of fasting.

Molting over and their appetites renewed, the first priority of the Inaccessible Island penguins was to restore their lost weight. The Stoltenhoff brothers reported that they came and went from the sea to feed for the next several months, until about mid-April—early in the Southern Hemisphere's autumn—when, as mentioned earlier, they abruptly disappeared. Where they went and what they did during the four months they were gone is unclear to this day. We know that they do not migrate to another island or distant land. Some scientists think they stay close to the Tristan group of islands, but there is suggestive evidence that they may venture far out into the ocean. For example, in the Southern Ocean, the *Challenger* scientists once sighted a group of penguins in the open sea, over two hundred miles from the closest island. And in 2016, scientists tagged penguins from New Zealand's sub-Antarctic Campbell Island at the end of their molting season, shortly before they took to the sea for the Antarctic winter. Months later, when they returned, the tags showed that they had ranged widely through the seas south of New Zealand and Australia, some swimming far south toward the Antarctic. Not once had they clambered up onto land. A few had traveled as many as nine thousand miles.

Recent studies have largely corroborated the Stoltenhoff brothers' observations about the annual rockhopper penguin cycle. They have also added many new details. For example, the brothers reported that most of the penguin nests contained two eggs. Modern researchers have confirmed the observation, but report that the first egg is always smaller than the second, and often it does not hatch. Even when it does, the first chick is invariably smaller than the second and rarely survives. The result is that for the most part only one chick is reared in each nest. Why do the penguins expend the energy to lay that first, unproductive egg? Nobody knows.

In order to track the penguins' underwater feeding behavior, researchers have used monitoring devices much like the tags attached to the New Zealand penguins. The record dive observed in these studies was to a depth of more than three hundred feet, and the longest underwater time recorded was eleven minutes. During a day of foraging the penguins normally spend more than a third of their time underwater, making many hundreds of dives, typically to depths of between ten or twenty and about a hundred feet in search of fish and other prey. The *Challenger* naturalists, too, were curious about the penguins' underwater behavior, but their experiments were much cruder than those of more recent researchers. George Campbell wrote that they put one of the rockhopper penguins in a lobster pot and "sank it a few feet." When they raised the pot after five minutes the penguin was dead. Campbell estimated that it had probably died in much less time; John Murray was more precise. Presumably referring to the same experiment, he wrote that a penguin submerged in a basket was "dead in one minute and thirty seconds," and that the birds' dives must therefore be much shorter than that. But Murray never took into account the likelihood that the poor penguin had suffered a panic attack when a strange, giant creature stuffed it into a basket and pushed it underwater.

The fossil record shows that penguins have been present on planet Earth for tens of millions of years. Thomas Huxley, who was an accomplished anatomist and an expert in bird anatomy, was the first person to describe a penguin fossil. When a colleague sent him a single fossil bone from New Zealand he immediately recognized it as a leg bone from a bird. In addition, its peculiar characteristics indicated that it came from a penguin. Huxley published his results in the *Journal of the Geological Society of London* in 1859. His most surprising conclusion, which was based on the size of the fossil bone, was that the ancient penguin must have been a giant. It would have been almost twice the size of the largest living variety, the emperor penguin.

The *Challenger* naturalists captured and preserved many adult and juvenile penguins, along with their eggs, for later examination. At the conclusion of the expedition all the specimens were sent to Morrison Watson, professor of anatomy at Owens College in Manchester (now the University of Manchester) for further analysis. Seven different species were represented in the collection, and Watson painstakingly dissected and described each one. His long, comprehensive report appeared in one of the Challenger Report volumes in 1883; in it he treats the penguins within the framework of Darwin's ideas about evolution. Comparing penguins from the *Challenger* collection with other types of birds, he concludes that penguins as a group must have "diverged at an early period from the primitive avian stem" that gave rise to all other varieties of birds. He cites Huxley's paper on the fossil penguin bone, but for want of further evidence from the fossil record he can say little about additional details of their evolution. "We are compelled," Watson wrote, "to postpone the accurate determination of the affinities of the [penguins] till the progress of Palaeontology shall have made us acquainted with the intermediate forms."

In the years since the 1880s, fossils of many of those intermediate forms have been discovered. Penguins, ancient and modern, tend to spend their lives in places that are not especially amenable to fossil preservation, and complete skeletons are quite rare. But penguin bones are dense and robust and distinctive enough that paleontologists have made good progress on the task of working out the affinities that so interested Morrison Watson. New Zealand, the source of Huxley's leg bone, continues to be one of the richest hunting grounds; Antarctica has also yielded many penguin fossils. The evidence from these finds corroborates Huxley's conclusion that his specimen came from a very large penguin. As it turns out, giant penguins, some almost human size, are abundant in the evolutionary history of these birds. And DNA sequencing has made it possible to work out relationships between penguins and other groups of birds in a way that was not possible from the physical evidence of the fossils alone.

The oldest penguin fossils currently known were found in New Zealand and are between 61 million and 62 million years old, indicating that penguins arose quite early in the history of birds. Two things are striking about these finds: first, the fossils show that at least two distinct varieties of penguin coexisted even at that distant time, and second, both species were substantially bigger than today's emperor penguins. In 2017, researchers described another giant fossil penguin, also from New Zealand, that is slightly younger than those ancient varieties. It is estimated to have weighed about 220 pounds and would have stood about five feet seven inches high. It is dated to around 57 million years ago and is distinct from its older relatives. So it appears that by 62–57 million years ago, multiple species of penguins had already evolved, some or perhaps most of them larger than today's largest penguins. This observation suggests that the answer to Morrison Watson's question about when penguins diverged from the "avian stem" is probably long before the 61–62 million-year age of the oldest fossils. The jury is still out on exactly when the divergence occurred, but some researchers have concluded that it must have been around 70 million years ago, or even earlier.

What was our planet like 70 million years ago? Certainly it was nothing like today's Earth: the global climate was very warm and the earth had no ice caps at the poles; North America was connected to northern Europe and the Atlantic Ocean as we know it did not exist; and Australia and New Zealand were still attached to the Antarctic continent. The Cretaceous period was hurtling toward its end, and reptiles still ruled: dinosaurs roamed the continents, and pterosaurs— those winged reptiles that are almost as popular as dinosaurs in museum displays and as children's toys—flew the sky. But birds, Watson's "avian stem," were present too. Fossil bird bones and skeletons from that time, as well as abundant fossil bird tracks, show that they were already diversifying. Several types of waterbird were present—fossil tracks of web-footed birds are fairly common—but to date no fossil penguins, or anything that could be considered an immediate penguin precursor, have been found from the Cretaceous. But 66 million years ago, everything changed. A large asteroid hit the earth, causing widespread devastation; the impact and the concurrent extinction of many groups of plants and animals, including dinosaurs and pterosaurs, mark the end of the Cretaceous period in the geological record. The mass extinction of living things opened up for new inhabitants numerous ecological niches that had formerly been occupied by Cretaceous plants and animals.

Birds survived the mass extinction event, and within a few million years an amazing array of carnivorous waterbirds had emerged, ancestors of birds ranging from today's flamingos to herons, pelicans . . . and penguins. This is not surprising. The oceans cover 70 percent of the earth's surface and contain an abundance of food; freshwater lakes and streams add to the total. After the extinction the early waterbirds diversified to fill the many newly available ecological niches. Some fed along the seashore, some in lagoons, others in freshwater. Some developed the ability to dive and swim underwater in pursuit of their prey. Penguins evolved to become completely marine. The "elbows" and "wrists" of what had once been wings fused, so that the wings became rigid flippers and powerful swimming paddles. Their body shape became streamlined, and their feathers flattened for efficient underwater travel. Their bones—which are light in most birds to facilitate flight—became denser, lowering the penguins' buoyancy and making it easier for them to dive deeply. And they became very big. They could not have flown if they wanted to; flying requires a low body weight, and the penguins were far too large. Adult loons, for example, close relatives of penguins, typically weigh about nine pounds. They are expert swimmers and divers and also excellent flyers—once they are in the air. But if you have ever watched a loon that has just filled up with fish trying to take off from a northern lake, you'll know that it requires a lengthy runway to get aloft. If it has overeaten it might not be able to take off at all. Emperor penguins weigh five or six times as much as a loon; the giant fossil penguins weighed much more. Once the ancestors of modern penguins gave up the skies and committed to a life at sea they no longer had to worry about their weight. But *their* ancestors must have been smaller birds that, like loons, were at home both in the air and the water. The evolutionary journey from those smaller ancestors to the enormous penguins that lived 62 million years ago probably took millions of years, stretching well back into the Cretaceous period.

The modern penguins of the Tristan da Cunha islands were the first encountered by Moseley and the other naturalists on board *Challenger*, but they were far from the last. After Tristan, the ship's next call was Cape Town, where she stayed for several weeks while being outfitted for her foray deep into the Southern Ocean. Moseley took the opportunity to explore the surrounding area, including a visit to a penguin rookery on a rocky outcrop in a bay not far from Cape Town. There he found a completely different variety of penguin. They were not rockhoppers; they waddled rather than hopped, and they did not leap in and out of the water like porpoises. They built their rudimentary nests on the bare rocks using whatever they could find: pebbles, shells, nails, bits of an old sail. They were commonly known as jackass penguins because they brayed like donkeys. Moseley considered them ugly compared to the rockhoppers of Tristan.

Of all the penguins he came across during the expedition, though, Moseley's favorite was the king penguin. He first saw these striking birds on Marion Island in the southern Indian Ocean, and he later encountered more of them on the Kerguelen Islands, also in the Southern Ocean. They were stunning, larger than any of the other penguins he had seen; they stood erect with their heads and bills pointing upward, showing off the bright slashes of yellow-orange on either side of their snow-white throats. Waddling about on two feet they looked almost like small people. Campbell's take on them was that they were pompous and solemn, befitting a king. He remained unimpressed with their habits on land: "They appear to pass their existence when on shore in standing still, yawning, occasionally pecking at their feathers, and sleeping—standing bolt upright, with their heads turned down on one side like an ordinary bird, only penguins have no wings to put them under. In this position they appear headless and eerie." Moseley was especially fascinated with the rookeries, which appeared to be divided into different sections for different purposes. In one section, hundreds of penguins would stand around, packed closely together, pompous and solemn. Another section was occupied only by breeding pairs, some still tending eggs, others with already hatched chicks. There were no nests. Eggs lay scattered on the bare rocks or enclosed, warm and snug, in a small pouch on the front of the bird, between its two webbed feet.

At Heard Island, another of the Southern Ocean islands where *Challenger* stopped, Moseley met a group of sealers. They had been there for many months, some for years, with only periodic visits by supply ships that brought them essential goods and took away the sealskins and oil they had accumulated. The island was desolate and inhospitable, and the men had little contact with the outside world. They ostensibly endured the isolation for the money. But for most of them financial security did not last long; they spent the bulk of their salaries at the first port they visited on the journey home. The sealers told Moseley that they ate penguins regularly. They saved the skin and its thick layer of insulating fat, dried it out, and used it as fuel for their fires. It burned well. Moseley's reaction is not recorded, but Campbell, who by then had tramped through many rookeries, had been nauseated by the stench, and had been ceaselessly bitten and pecked, was pleased. "An admirable way of making these fiends of some use," he wrote. Neither man said anything about the foul smell this practice must have created in the sealers' houses, which were little more than small, roofed huts erected over a hole in the ground.

The islands on which the *Challenger* naturalists encountered penguins are volcanic, and most are active (that is, have erupted sometime during the past several thousand years and may erupt again). Throughout the Southern Hemisphere penguin colonies are often situated on or close to active volcanoes. Surprisingly, though, until recently no one had investigated how this natural hazard might affect penguin populations or evolution. But a study published by an international group of scientists in 2017 suggests that the impact is large.

The researchers examined long-term population trends in a penguin colony on a small island that is part of the Antarctic Peninsula. The island is also home to a small but long-lived lake from which the researchers were able to take sediment cores; analysis of the cores enabled them to chart changes in the penguin population over time, reaching back about seven thousand years. Several times during that period, the measurements revealed, penguin numbers had suffered rapid, nearly catastrophic declines. Recovery after each of these episodes took many hundreds of years. Every one of the declines coincided with a period of eruptions from a nearby volcano that had spread layers of volcanic ash throughout the region.

No other environmental change the scientists could measure—differences in the amount of sea ice, for example, or fluctuations in sea or air temperature—could account for the results. The only phenomenon that correlated closely with the population changes was volcanic eruptions. The researchers speculate that the penguins would have been adversely affected both directly and indirectly by the volcanism, which would have resulted in burial of chicks and eggs in ash, ingestion of the tiny abrasive shards of volcanic glass that make up the bulk of the ash, loss of prey, and even tsunamis or mudflows initiated by rapid melting of glaciers that had been covered with dark ash.

Even active volcanos can experience long periods of quiescence, and not all eruptions are devastating, to penguins or other lifeforms. Tristan da Cunha is an active volcano, but until 1961 its human inhabitants had thought it was perfectly safe. Then in October of that year, after several months of small earthquakes, the volcano started to erupt. Alarmed residents watched as a dome of lava pushed up on the flanks of the main volcanic peak, very close to the village of Edinburgh of the Seven Seas. At night the red glow was readily visible, and the villagers made a collective decision to evacuate to a safer part of the island. After one harrowing night, however, they decided to leave the island entirely, and they evacuated again, this time by boat, to nearby Nightingale Island. The situation seemed serious enough that a passing ship was diverted to rescue the islanders, who were taken first to Cape Town and then to England. It was two years before most of them would return to Tristan da Cunha.

But the islanders were lucky. On the scale of volcanic events, the Tristan eruption was minor. Lava engulfed the canning factory in the village and flowed onto the beaches that had been used as landing spots for the fishing boats, but there were no casualties. In geologic parlance this was a "parasitic" eruption, a new, small volcanic cone that formed on the side of the parent volcano. And it was not explosive like the eruptions that blasted tons of volcanic ash into the atmosphere and affected penguin populations on the Antarctic Peninsula. A small amount of ash was deposited locally on Tristan da Cunha near the new volcanic cone, but the island's flora and fauna survived largely unscathed. Moseley's penguins, *Eudyptes moseleyi*, suffered no serious consequences. On Tristan da Cunha, harvesting of penguins and their eggs by residents in the past, depletion of the penguins' food supply by commercial fishing, and the introduction of nonnative animals such as dogs, cats, pigs, and rats—the human pressure of just a few hundred people, extended over the 150 years or so that the islands have been inhabited—have had a much greater effect on penguin populations than has any natural phenomenon. With a little luck the endangered status of northern rockhoppers and the newfound ethos of penguin conservation among the islanders will reverse the trend.

Heard Island, where the sealers burned penguin hides as fuel, lies at 53° south latitude, more than thirty-six hundred miles, as the crow flies, south of the equator, and about twenty-five hundred miles from the South Pole. It has permanent glaciers; when *Challenger* visited at the height of the Southern Hemisphere summer, the naturalists described the glaciers as descending to sea level and being washed by the waves. In North America the fifty-third parallel of latitude passes through the wheat fields of the Canadian prairies, and in Europe it crosses England about a hundred miles north of London, through a landscape that features green fields, not glaciers. The difference arises because the Gulf Stream carries heat from the tropics far north in the Atlantic, warming Europe; in the Southern Hemisphere no such current exists. Instead, in the Southern Ocean, the broad swath of the Antarctic Circumpolar Current circles the globe from west to east around Antarctica, carrying more water than any other ocean current on the planet and creating a barrier to the transport of heat farther south. Its waters are cold, and they refrigerate places like Heard Island.

Challenger left Cape Town in mid-December 1873 and headed south toward the Antarctic, visiting a number of sub-Antarctic islands along the way. Heard was the last; it would be months before the scientists saw land again, when they reached Australia. The timing of this stage of the expedition had been planned carefully; it was February, the optimum time of year for exploring the Southern Ocean. Little was known about the region. A number of explorers had ventured south of the Antarctic Circle before, and several had sighted land, but only two (British explorer Sir James Ross, and Frenchman Jules Dumont d'Urville) had managed to land on the Antarctic continent—the size and nature of which were wholly unknown. The official report of the *Challenger* expedition summarized the existing knowledge: land in the Antarctic was covered in ice and snow, was in places mountainous, and was fringed along the coast by an "ice barrier." (Ross had reported sailing 450 miles along a precipitous wall of ice almost 200 feet high.) It was unclear whether the land was continuous—a real continent—or made up of a series of islands. However, no one on *Challenger* anticipated getting far enough south to set foot on land and add to the fragmentary existing knowledge, much as the naturalists would have loved to do so. Cognizant of the dangers of ice, when they passed the Antarctic Circle and saw ahead of them large icebergs stretching as far as the eye could see, they pulled back. The Challenger Report explains that their goal "was not to obtain a particularly high latitude, but merely to make observations on the temperature and depth of the sea in the vicinity of the ice." Both scientists and naval personnel realized that "it would have been foolish to go further south in an unfortified ship with only six months' provision on board."

Their caution was not misplaced. Even in the Antarctic summer, the Southern Ocean is an inhospitable place. The *Challenger* crew got a taste of its fury shortly after leaving Heard Island, still far north of their deepest penetration toward Antarctica, when a huge wave broke through several ports on the side of the ship, "floating everything out of the sick bay." Later, when they were farther south and among icebergs, the weather improved. They experienced several calm and sunny days that allowed all on board to appreciate the grandeur and beauty of the ice. But the benign weather was followed by days of cloud, mist, and blowing snow that reduced visibility almost to zero and made the going slow. In such conditions everyone had to be on high alert for looming icebergs, especially during the few short hours of darkness. Twice they were engulfed in howling gales, forcing *Challenger* to shelter in the lee of icebergs that towered over the ship and shifted and moved with the wind and currents. They had to maneuver constantly to avoid collision. It gave them, as the understated Challenger Report notes, several "anxious nights." Once, navigating in heavy winds close to a massive iceberg, they collided with it. The jib (the front sail that projects forward from the bow) was completely sheared off. It was definitely an anxious moment. Repairing the damage in the intense cold and wind was not easy, and in the midst of the work another iceberg materialized out of the mist, initiating a frantic burst of activity—engines thrust into reverse at full throttle, officers shouting orders over the noise of the storm, crew working the sails to prevent another collision. A seaman fell from the mast and was badly injured. Finally *Challenger* cleared the berg. A "narrow shave," wrote George Campbell. "An iceberg in fine weather is a beautiful sight, but in a fog and gale of wind it is very much the reverse."

In today's era of omnipresent communication it is not easy to imagine the stark isolation of the *Challenger* expedition in the Southern Ocean. Her crew had encountered a few sealers and whalers at Kerguelen, and they met more sealers on Heard Island, but their last point of contact, as far as anyone in Britain was concerned, had been Cape Town. That was in mid-December; it was now February. They were alone in the vast expanse of the Southern Ocean, a tiny speck among icebergs that were sometimes hundreds of feet high and miles long. They were not due to reach Melbourne, Australia, until mid-March. In the meantime, nobody would know anything about their whereabouts or their condition. If the ship collided with an iceberg and was damaged beyond repair, no one would be any the wiser until she failed to show up in Melbourne. Even then it would be impossible to know what had happened, or where to start looking for her crew; they would be presumed lost at sea. Did those on board think about such dangers in these terms, especially during their "narrow shaves"? Nothing in the official records of the expedition addresses this question, but it is likely that they felt fear at times. They were well aware of the hundreds of men, mostly whalers but also some explorers, and the many ships that had been lost to ice and weather in the Arctic. Most of those ships had been much better equipped for polar work than *Challenger*. But in another sense the harsh conditions were the crew's reality at that moment and they simply continued with their normal daily routines. The sailors and officers were intent on bringing the ship safely through any difficulties, and the naturalists were focused on their science. They endured the storms and went about their tasks of navigation, sounding, trawling, taking temperature measurements, dredging, and recording observations much as they had done during less perilous parts of the voyage. The accounts of the expedition penned by several of the participants—Henry Moseley, George Campbell, Joseph Matkin—as well as the official Challenger Report all mention the dangers, but do not dwell on them.

Icebergs were known to the *Challenger* naturalists from the descriptions of other explorers, if not from personal experience. Wyville Thomson and John Murray may have seen some during their travels in the North Atlantic. But still they were objects of intense interest and anticipation, not just for the scientists but for almost everyone on board. (After their several narrow shaves, though, most of the seamen had seen enough and were happy when the ship began sailing north again, away from the ice.) Starting with the first iceberg they spotted, on February 10, 1874, the naturalists numbered each one and noted down its position, as well as its physical characteristics: shape, size, color, anything that seemed peculiar. Before long, however, they had to abandon their scheme. With tens and sometimes hundreds of icebergs within sight of the ship at any one time there were far too many to record individually. But the beauty of the floes, especially in good weather, astonished all on board, as is obvious from their journals:

> No words of mine can describe the beauty of these huge icebergs—one, which we have just sailed past, had three high caverns penetrating a long way in; another was pierced by a hole through which we could see the horizon; and the wonderful colouring of those blue caverns, of the white cliffs, dashed with pale sea-green, and stratified with thin blue lines veining the semi-transparent wall of ice 200 feet in height! [George Campbell]

> The colouring of the southern bergs is magnificent. . . . There are parallel streaks of cobalt blue, of various intensities. . . . The colouring of the crevasses, caves, and hollows is of the deepest and purest possible

azure blue. None of our artists on board were able to approach a representation of its intensity. [Henry Moseley]

We passed one [iceberg] on the night of the 14th, which looked something like Windsor Castle, & had a large cavern through it which showed the daylight on the other side; nearly all hands were on deck to see it, & one of the men even said, he had often paid a penny in England to see not half such a sight. [Joseph Matkin]

No painting could realize it, and if it could, you would not believe its truth; the colour and exquisite softness of the blue, from light azure to indigo in successive shades as the cavern penetrated deeper and deeper into the berg, with fringes of icicles hanging from the roof. [George Campbell]

Henry Moseley, as we have seen, had interests that ranged far beyond his primary duties as a biologist. In the Southern Ocean he sketched and described icebergs, and tried to work out for himself the origins of their various features. He noted that most were rectangular and flat-topped, and surmised that they had broken off the ice barrier that, according to James Ross and other explorers, fringed the land of the Antarctic. Moseley's basic conclusion was correct. The glaciers of the massive Antarctic ice cap slowly flow outward from the center of the continent toward its edges, and in some places they slide directly into the ocean. If a glacier pushes out into water deep enough for buoyancy to take over, the ice floats off the rocky bottom, forming an ice shelf. As it continues to move farther and farther out to sea, warm seawater—barely above the freezing point, but still warmer than the ice—melts and corrodes it from below. Eventually an iceberg breaks off; typically the ones formed in this way are flat-topped and roughly rectangular, as Moseley described. They can be huge. The *Challenger* scientists estimated that the largest one they encountered was about three miles long. In 1956 an American icebreaker, U.S.S. *Glacier*, reportedly spotted one that was approximately 208 miles long and 60 miles across. And in 2000, with the advantage of satellites, scientists watched the Ross Ice Shelf—named after James Ross—crack and break off what is so far the biggest iceberg ever precisely measured. It was smaller than the one reported by *Glacier*, but still gigantic: 183 miles long and 23 miles wide and estimated to weigh over three trillion tons. Its wanderings through the Southern Ocean were tracked for five years as it gradually melted and broke up. Satellite monitoring now permits scientists to routinely detect large icebergs breaking off Antarctic glaciers, and although none as large as the one seen in 2000 have been observed since, in recent years the rate of Antarctic iceberg formation appears to have speeded up—most likely as a consequence of global warming.

Moseley also sketched and described icebergs with more complicated shapes than the typical flat-topped, rectangular variety, and he tried to understand their origins. Some had peaks and pinnacles, some had multiple, steplike "stories," others were riddled with caves and arches. All these features, he reasoned, were connected with the iceberg's age. Bergs freshly broken off an ice shelf were gradually worn down by the force of the sea constantly beating against them; "wash lines" developed near sea level, where the waves ate into the ice. When the overhangs created in this way could no longer be supported, they collapsed. Now lighter, the iceberg would float higher, and a new wash line would begin to form. Constant battering by the waves formed caves in places where the ice was weak because of a pre-existing crack or crevasse. The sound of waves sweeping into the caves and smashing against the ice was thunderous, and the pounding enlarged the caves—sometimes pushing them right through a narrow iceberg to form an arch. When an arch collapsed, it left pinnacles. Moseley writes about all these phenomena with the same enthusiasm and analytical reasoning that he displays when he discusses his biological investigations. Taken together, he said, these iceberg features reminded him of a coastal landscape. He had found an analogy between the erosion of icebergs and the weathering action of the sea on a rocky coastline.

He was especially intrigued by the prominent stratification that was present in most of the icebergs. The naturalists had hoped to land on one of the bergs for closer study, but that proved impossible, so the best Moseley could do was examine nearby ice with a hand-held telescope. He observed that the deepest layers nearly always appeared

One of the older, more complex icebergs that the *Challenger* scientists observed, this one tilted, with wash lines, caverns, and overhanging ledges. It was spotted on February 19, 1874, at 64°37′ south latitude. (Sketch by J. J. Wild, from *Report on the Scientific Results of the Voyage of H.M.S. Challenger . . . Narrative; Volume 1, First Part*, plate C, fig. 5.)

to be thinner than those above, and he assumed this was because they had been compressed by the weight of the overlying ice. This, he thought, must continue in the deeper parts of the icebergs, the main mass that was below sea level and hidden from view. In most of the bergs the layering was perfectly horizontal, but in a few—typically the old eroded bergs with complex shapes—it was tilted, an indication that the iceberg had become unbalanced through erosion, and its center of mass had shifted. As an experiment, the scientists once fired the ship's twelve-pound cannon at a nearby iceberg. The cannonball thudded into the ice, breaking off part of a cliff, which slid into the sea and filled the area around the ship with chunks of ice; the scientists quickly retrieved as many fragments as they could. Moseley found that the layers he had until then seen only from afar were made up of alternating bands of hard, transparent ice and whitish, less compact bands. At a distance, the transparent ice took on a blue hue, which gave the icebergs their dazzling blue coloration. The layers eroded differently; Moseley wrote that this gave the appearance of the leaves of a book when viewed from the ship, the softer whitish bands slightly eroded by the waves, the harder transparent bands less affected.

Moseley was certainly not the first to be interested in glacial ice; many other scientists had studied the ice of European glaciers, and to a lesser extent Arctic ice, at close range, and they had an understanding of how it forms. But since the time of the *Challenger* expedition glaciologists have learned much more about ice in the Antarctic. The layers Moseley observed are a record of annual deposition: the hard transparent bands—as he had deduced—caused by summer melting and refreezing, the white, more porous bands a layer of unmelted winter snow. As the layers build up year after year the lower ones become compressed, air in the pore spaces is squeezed out, and eventually the ice recrystallizes into a dense, solid mass. Because ice efficiently absorbs most colors of the spectrum except blue, large pieces of dense glacial ice appear intensely blue. (Liquid water appears blue for the same reason.)

The icebergs that *Challenger* sailed among were just tiny fragments of the vast expanse of glacial ice that covers the Antarctic continent. At the time, no one knew its extent or thickness. But the annual layers that Moseley described have proven to be a godsend for scientists. In the thickest parts of the icecap these layers have piled up continuously for around eight hundred thousand years, and by drilling in such areas glaciologists have collected ice cores that contain a layer-by-layer record of the snow that has fallen on the southern continent over that entire time. Moseley's description of the layers resembling pages of a book is apt; written within them is eight hundred thousand years of Antarctic environmental history. It takes ingenuity to decipher the record, but with the right kind of measurements everything from the amount of carbon dioxide in the air in the distant past to the temperature, the strength of the wind, and the occurrence of volcanic eruptions can be read from those pages. And recently a remarkable find was made: Antarctic ice that is 2.7 *million* years old. A group of researchers studying glacial flow in the icecap realized that in some places, deeply buried ancient ice flowing outward toward the coast might have encountered a snow-and-ice-covered mountain range and been forced up toward the surface. These scientists spent two field seasons taking ice cores in one of these regions, and their theory proved to be correct. The cores they retrieved sample only isolated parts of Antarctic history and do not provide a continuous record from the present back to 2.7 million years, but the ice they *do* sample has the potential to illuminate a far more distant period than anyone thought was possible even a few years ago. Among other things, 2.7 million years ago is about the time when the great ice sheets of the northern hemisphere, which eventually reached down into the United States and much of Europe, were just beginning to form. What can the Antarctic cores reveal about that episode? The hunt is now on for drilling sites that might lead to even older ice. Moseley realized that the strata he saw so clearly in the icebergs, like the strata of sedimentary rocks, represented slices of past time. But he could not have imagined how far back into the history of the Antarctic they would reach.

When *Challenger* left England on December 21, 1872, the scientific consensus was that the floor of the ocean was covered with a thick layer of calcareous mud made up mostly of the shells of organisms such as the foraminifer *Globigerina*, as we saw in Chapter 4. The discovery of red clay covering large parts of the Atlantic seafloor disrupted that theory. Yet another surprise was in store for the *Challenger* scientists in the Southern Ocean.

As the ship sailed among the icebergs and pack ice, the naturalists went about their regular activities as weather permitted. When he was not sketching icebergs, Moseley went off in a small boat to shoot birds to add to the *Challenger* collection. (I'll discuss the Victorian obsession with collections in Chapter 10.) The scientists and the ship's crew regularly took depth soundings and dredged the bottom. On February 11, 1874, just a day after sighting their first iceberg, *Challenger* stopped for an official observing station (station 152) at approximately 61° south latitude. The depth was measured and serial temperature readings were taken through the water column. Then the scientists trawled the seafloor. The depth was 7,560 feet, and the trawl came up with something they had seen only once before on the voyage: fine-grained mud that was superficially similar to the familiar

Globigerina ooze of the Atlantic, except that this mud was composed of tiny shells made of silica, not calcium carbonate. The siliceous shells were the remains of small planktonic organisms called diatoms, a variety of algae, and the station log identifies the sediment as diatom ooze. *Challenger's* single previous encounter with this kind of sediment had been about six weeks earlier on the northern fringes of the Southern Ocean, near Marion Island.

Diatom shells appeared in dredge hauls and net tows throughout the expedition, but nearly always in low abundance—perhaps 1–2 percent of the total. At several stations in the Southern Ocean, however, they dominated. The *Challenger* scientists were not entirely surprised; Joseph Hooker, the naturalist who had accompanied Ross to the Antarctic more than thirty years earlier, described the Southern Ocean as "swarming" with diatoms. They were so abundant, he noted, that in places they gave the water a brownish tint and even discolored the icebergs. A few days after they dredged up diatom ooze at station 152, Moseley reported that nearby icebergs were stained an "ochre tint" by diatoms, and he took one of *Challenger's* small boats out to collect pieces so that he could examine the phenomenon more closely.

Sediments are rarely, if ever, pure, and designating the sediment type at each of *Challenger's* sampling stations was done by estimating the abundances of various components and deciding which was dominant: calcareous ooze, red clay, diatom ooze, and so on. Many of the dredge hauls in the Southern Ocean yielded diatom ooze, but even when diatom shells were not dominant they were still abundant. As more and more dredging stations were occupied a coherent picture began to emerge. At the southernmost stations, closest to the pack ice and presumably to the land beyond, the main component of the sediment was something Murray called "blue mud." It contained grains of clay, quartz, and other minerals, and it had all the signs of originating from land. It was similar to the sediments dredged off the coasts of Europe, the Americas, and Africa. Even at the blue-mud stations, however, significant quantities of diatom shells appeared. At more northerly stations, slightly farther away from the presumed land, the sediment was diatom ooze—nearly always with a little blue mud mixed in. It looked as though there were two bands of bottom sediment, roughly parallel to the lines of latitude: blue mud in the south and diatom ooze in the north. One type of mud transitioned into the other without a distinct dividing line. The ubiquitous diatom shells obviously came from the organisms in the surface water; the blue mud was there because somewhere to the south, the land of the Antarctic fed its erosion products into the ocean. Sometimes the blue mud contained pebbles and rocks in addition to clay and fine particles, and some of the larger pebbles showed the unmistakable signs of polishing and scratching by glaciers. Some were also rock types typical of continents rather than volcanic ocean islands. Together these observations were a clear indication to the *Challenger* scientists that substantial tracts of glacier covered land were present to the south, and not very far off. But such was the prevailing uncertainty about the nature of this remote part of the world that the section of the Challenger Report describing the blue mud reads, "The nature of the rock fragments dredged in these latitudes *conclusively proves the existence of continental land* of considerable extent within the Antarctic Circle" (italics added). That it was felt necessary to include such a sentence is another reminder of how little was known of the southern continent in the 1870s. The *Challenger* scientists never spotted the Antarctic land they inferred from the blue mud. Today, cruise ships regularly take tourists to the same latitudes and beyond, and sometimes even land them on Antarctica to walk with penguins.

The *Challenger* dredges delineated the bands of blue mud and diatom ooze across only a small part of the Southern Ocean. Murray, however, extrapolated far beyond the region surveyed by the ship. He knew that Joseph Hooker's report of diatom-rich seas referred to a different part of the Southern Ocean, but at about the same latitude as the locations where *Challenger* dredged diatom ooze. The seafloor under the region Hooker described would presumably be covered in diatom ooze, too, and Murray concluded that the siliceous sediments probably stretched right around the entire Antarctic polar region—a band of diatom ooze circling the earth like a halo.

Not everyone agreed. The Italian scientist who eventually examined the expedition's diatom collection and wrote the section of the Challenger Report dealing with it considered Murray's conclusion premature. If it were true, he reasoned, a similar band of diatom ooze should show up at high latitudes in the North Atlantic, where the climate was somewhat similar. Yet none had been found. Subsequent research has proven Murray correct. There *is* a continuous swath of diatom ooze circling the whole of Antarctica. But it is there because of specific oceanographic conditions, not just latitude or temperature, which is why no equivalent feature exists in the North Atlantic.

Diatoms are present almost everywhere in ocean surface waters, a fact that explains why they were found in small numbers in many *Challenger* dredge hauls far from the Antarctic. But diatom ooze accumulates only in places where the organisms occur in such massive numbers that their shells outweigh everything else that falls onto the seafloor. That can occur only in regions where the surface water contains both abundant nutrients—the essential components of life—and enough silica for building the diatoms' shells.

The most prolific source of both these components is nutrient-rich water that upwells from the deep ocean. Why is the upwelling water rich in nutrients? In the simplest terms it is because sinking particles slowly dissolve as they descend toward the seafloor, releasing nutrients, silica, and various other constituents. Deep currents, flowing through all the oceans, accumulate these components. The oceanic circulation system has been described as a conveyor belt, an apt metaphor: the currents incorporate whatever rains down on them from above; the farther they flow, the richer they become in these constituents. Wherever this nutrient-rich deep water upwells, it fuels biological growth at the surface. Nearby land can also be a source of nutrients and silica, washed into the ocean by weathering of a continent or an island. Around the Antarctic both these sources are present— the glaciers that scour the Antarctic continent carry a supply of silica-rich, easily dissolved "rock flour" into the Southern Ocean, and because of the way ocean currents work, large quantities of nutrient-rich deep water upwell toward the surface in the region. The result is a rich ecosystem that teems with life of all kinds. Diatoms play a key role because they are a primary food source for the famous Antarctic krill, small shrimplike animals that are themselves a main source of food for larger fish, squid, seals, penguins, whales, and others. In terms of their sheer weight krill are probably the most abundant animals on earth, so you can imagine the huge number of diatoms necessary to support them. No wonder the seafloor is blanketed in diatom ooze! The numbers are so vast that, as photosynthesizing organisms, diatoms play an outsized role in the production of the oxygen we breathe. The impact of diatoms worldwide is estimated to be comparable to that of all the world's rain forests combined.

Sedimentary deposits that are rich in fossil diatoms occur on land in a number of places throughout the world. Referred to as diatomite, or sometimes diatomaceous earth, this material is used in an array of commercial products, from cat litter to nail polish. The porous structure of diatomite makes it ideal for filters; if you drink beer or wine, chances are it has been filtered through fossil diatoms.

Diatoms are not the only marine organisms that secrete siliceous shells and skeletons. Radiolarians, another variety of tiny, floating plankton, do too, and in a few places where they are abundant in ocean surface waters their skeletons dominate the bottom sediments. In the equatorial Pacific, long after *Challenger* had left the diatom-rich Southern Ocean, the scientists again dredged up siliceous sediment, but this time it was radiolarian ooze, not diatom ooze. Like diatoms, radiolarian remains had been present in minor amounts in many *Challenger* dredge hauls from different parts of the oceans. But here they were the main component, and, again as with the diatoms, the reason had to do with upwelling. Near the equator the prevailing winds cause surface ocean water to move in opposite directions in the two hemispheres, essentially parting the sea and allowing cool, nutrient-rich water to well up from below and take its place. Radiolarians flourish in this environment, and their siliceous skeletons sink to the seafloor to become the radiolarian ooze. By the end of the voyage, the *Challenger* collection included 4,318 different species of

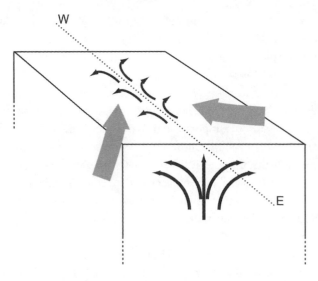

On a rotating earth, fluid movement is influenced by the Coriolis effect, which causes ocean currents to veer away from their original direction of travel. Near the equator, the converging trade winds (the thick gray arrows in the diagram) blowing from the northeast and southeast push surface water westward. But because of the Coriolis effect these currents diverge away from the equator—to the right in the Northern Hemisphere, to the left in the Southern. Nutrient-rich water from deeper regions upwells to take the place of the water moving away. In the Southern Ocean the upwelling has a slightly different origin but is controlled by the same principles: there the prevailing westerly winds push surface water toward the east, and it veers left, away from the continent, creating space for nutrient-rich water to upwell.

radiolarians, most of them from the nutrient-rich areas of the equatorial Pacific. More than 80 percent—a total of 3,508 species—were new to science. Most radiolarians live near the sea surface, but the *Challenger* dredges also brought up some that lived on or near the seafloor. One group was entirely new to science, and Murray gave them the biological family name Challengerida in honor of the expedition. Within this family he identified a number of separate genera, and again he named one of these after the *Challenger*: the genus

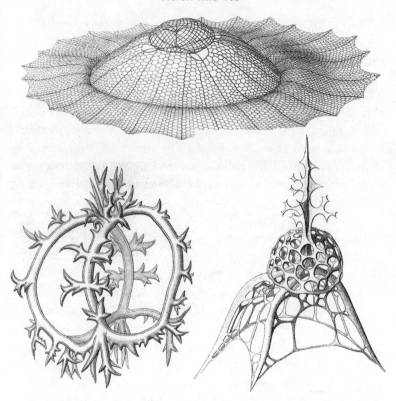

A few of the many intricately shaped radiolarians illustrated in Ernst Haeckel's monograph in the Challenger Report. Highly magnified here, they range in size from about two one-hundredths of an inch across for the specimen at the top to a less than half that for the bottom two. (*Report on the Scientific Results of the Voyage of H.M.S. Challenger . . . Zoology; Volume 18, Plates* [London: Her Majesty's Stationery Office, 1887], [*top*]: plate 70, fig. 1; [*bottom left*]: plate 82, fig. 8; [*bottom right*]: plate 60, fig. 10.)

Challengeria. Various species of *Challengeria* are named after the naval officers on the ship: *Challengeria nareii* after Captain Nares, *Challengeria campbelli* after George Campbell, and so on. There is also a *Challengeria murrayi.*

Responsibility for examining the radiolarian collection was given to Ernst Haeckel, at that time the world's foremost authority on these organisms (it was Haeckel who identified and named *Challengeria murrayi*). His study took much longer than expected. According to Murray, Haeckel devoted a decade of his life to the work, and the end result was a landmark study that not only analyzed the *Challenger* collection but also summarized all known information about the organisms. It ran to over 2,100 pages of text and included 141 plates with detailed drawings of microscopic, elaborately decorated radiolarian skeletons. Haeckel was especially interested in patterns and symmetry in nature, and radiolarians have both in abundance. Later in his life, at the age of seventy, he published a two-volume set of prints he called *Artforms in Nature*, in which intricate drawings of radiolarians figure prominently.

The radiolarian ooze from the equatorial Pacific rounded out the *Challenger* scientists' list of major ocean sediment types: calcareous *Globigerina* ooze, red clay, siliceous ooze dominated by either diatoms or radiolarians, and the blue or green clay found close to the continents. These results were quite different from what they had expected at the onset of the expedition, when they assumed that the ocean floor would be entirely covered in a layer of *Globigerina* ooze. By the time the section of the Challenger Report on deep sea sediments was completed, fifteen years after the expedition returned to Britain, Murray and his colleague Alphonse Renard were able to put together a comprehensive picture of how sediments are distributed on the seafloor. They had had the advantage of examining not only the *Challenger* samples but also sediments from regions not visited by *Challenger*, sent to them by colleagues from around the world. Even so, their synthesis of this material is little short of remarkable. In its rough outlines, their map of seafloor sediment distribution is not much different from the maps in oceanography textbooks today. It was an extraordinary achievement. Even in a three-and-a-half-year voyage *Challenger* could not possibly visit every inch of the world's oceans, and the sampling frequency along her track, though impressive for the time, was not particularly high. The samples sent to Murray and Renard from colleagues helped fill in some gaps, but still they had a limited database to work with, and it is somewhat surprising that they were able to map out sediment types as accurately as they did. Today, with the benefit of vastly more detailed sampling, we know much more about the distribution of sediments on the seafloor than Murray and Renard did, but in spite of that the basic elements they outlined have not much changed.

Once it became clear that different sediment types occupied different parts of the ocean floor, the question was, Why? Obviously, for the calcareous and siliceous oozes, part of the answer had to do with the types of organisms that flourished in the sunlit surface waters. Another key parameter was depth. When plankton living near the surface die, their shells and skeletons begin to dissolve as they sink toward the bottom. The farther they have to travel, the more they dissolve; some species are more susceptible to dissolution than others, and the substance of the shells—calcium carbonate versus silica, for example—also plays a part. Calcareous sediments are never found in the deepest parts of the oceans because the shells dissolve completely before they reach bottom. In the Challenger Report, Murray wrote about the sediment collection: "We frequently requested our assistants to select for us a sample from among several thousand, but not to give the slightest indication of the ocean or depth from which the specimen was obtained. After examination, we have then marked regions on the chart in which we believed it was collected stating at the time the probable depth. In the great majority of cases, in about nine out of ten trials, the position could be stated within a few hundred miles and the depth within a few hundred fathoms." Murray was not trying to impress his readers; rather he was reflecting on the fact that *Challenger*'s sampling program had revealed a logical order in the way the major, recognizable varieties of ocean sediment are distributed on the seafloor. This newly gained knowledge allowed him to pinpoint the probable location and depth of a random sample simply by examining it.

Impressive as that insight is in the sense of extending our general knowledge about the planet, there is more. The sediment distribution Murray and his colleagues on *Challenger* mapped out is partly a consequence of the great global geochemical cycles that operate on the earth. The calcareous sediments like *Globigerina* ooze are particularly important because they involve carbon dioxide, one of the greenhouse gases responsible for global warming. Although multiple chemical steps are involved, the calcareous oozes on the seafloor, and their ancient equivalents on land (limestone formations like "the chalk"), can be thought of as storehouses of carbon dioxide extracted from past atmospheres. The carbon they contain ultimately came from atmospheric carbon dioxide dissolved in seawater and taken up by the planktonic organisms; if it were somehow released again all at once the earth would be like Venus, with a carbon dioxide–rich atmosphere and surface temperatures twice as high as the highest setting on your oven.

Challenger's dredges only scooped up the topmost layers of seafloor sediment, down to depths measured in inches, and therefore the pattern of sediment distribution Murray and his colleagues mapped out reflects the present-day state of the oceans. But in places the blanket of sediment is miles thick. What if it were possible to sample those deeper layers? They were laid down in the distant past; would they be the same as those being deposited today at the same location or would they be different, the product of different ocean conditions in the past? Questions like these have driven oceanographers to develop ways of sampling deeper into the seafloor mud than was possible on *Challenger*, starting with simple gravity-driven pipes that plunged into the sediment and eventually evolving to sophisticated deep sea drilling rigs on dedicated ships that can bring back cores a mile long. One of the first sets of long sediment cores ever recovered was collected by the Swedish Deep Sea Expedition of 1947–1948. The scientists on this expedition used a newly developed coring device that in favorable cases could retrieve cores sixty or seventy feet long.

The Swedish expedition spent much of its fifteen-month voyage in the Pacific and by design occupied several of the same stations that *Challenger* had sampled some seventy years earlier. The results from these places confirmed the sediment types reported by the *Challenger* expedition. But more important, the long cores allowed the Swedish scientists to look back in time, to address the question of whether conditions had been different in the past. What they discovered was that there had indeed been changes. Radical changes. There were, the scientists noted, "highly interesting stratifications" in the cores. Sometimes the type of sediment present changed completely over a small section of a core—for example, from *Globigerina* ooze to red clay and back again. The Swedish scientists tentatively ascribed these changes to variations in climate that had affected ocean conditions. They suspected that the alternating sediment types might be related to ice ages.

Today deep sea sediment cores are one of the cornerstones of research into the earth's climate history back to almost 200 million years ago, the age of the oldest sediments that can be found on the seafloor. As the scientists on the Swedish Deep Sea expedition anticipated, sediment cores have been particularly useful for working out the timing and extent of temperature changes through the warm/cold cycles of the Pleistocene Ice Age of the past three million years or so. Those cycles were accompanied by changes in the carbon dioxide content of the atmosphere—cold periods had less, warm periods had more—and the fluctuations are reflected in the amount of calcareous *Globigerina* ooze in the sediments at particular locations and ocean depths. Such progress would have been beyond the imagination of the *Challenger* scientists. But their delineation of the major ocean sediment types, and the hints that climate might have something to do with how they are distributed on the seafloor, laid the foundation for the work that followed.

On March 4, 1874, the crew of *Challenger* said good-bye to the last Antarctic iceberg they would see on the voyage. They were en route to Australia and the Pacific, and no one was particularly sorry to leave the cold seas of the Southern Ocean behind. Two weeks later the ship anchored at Melbourne, and George Campbell described their arrival after a long sojourn at sea with his characteristic succinctness and wit: "There was joy among us on arriving at Melbourne. Of gales, snow, icebergs and discomfort generally, we had had enough, and the memory of a dinner I ate at the club the first evening, followed by the opera, yet lingers in my memory as one of the pleasantest experiences of a poorly paid and laborious career!" Still, the experiences in the Southern Ocean had not all been negative. Aside from the science accomplished, most on board relished the personal satisfaction of having seen things that few others had. Campbell went on to say that the voyage was worth the discomfort, for few people then living had ever had the pleasure of enjoying such spectacular iceberg scenery.

Challenger would spend most of the next two years in the Pacific. Her remit was to explore the deep ocean; Wyville Thomson later explained that he "rather discouraged" too much effort by the naturalists on land, especially if it took away from work at sea. But particularly for Henry Moseley, islands were like magnets. In the Pacific, *Challenger* would visit many islands that, while perhaps not as remote as some of those she had explored in the Atlantic and Southern Oceans, were much more exotic. They had an additional attraction: native people. Anthropologists—most of them European—had a fascination with people who were different from themselves, especially those they considered to be "primitive." The scientists of the *Challenger* expedition shared that interest and that prejudice. Along with the fish, coral, insects, birds, rocks, and other natural specimens they collected, they brought back for further study tools, weapons, and decorative articles made by native people. They also collected skulls and partial skeletons of indigenous people from South Africa, South America, Australia, New Zealand, and various Pacific Islands; the human remains were later examined by anatomists, and their findings were written up in minute detail in the Challenger Report.

The context in which the *Challenger* scientists worked was vastly different from that of today, perhaps most strikingly in their interactions with indigenous people. The conventional wisdom at the time—at least among those in what we today refer to as "the West"—was that European civilization was superior to others, and probably always would be, a belief now referred to as cultural imperialism. Moseley's treatment of native peoples, for example, was not much different from his approach to corals or penguins: the people, their customs, and their tools and weapons were objects of study, to be measured, described, and perhaps displayed in a museum. It is unsettling to come across frequent references to indigenous people as "savages" in his book about the expedition. He was not alone, however; the term appears in the writing of other scientists and ship's officers about the voyage, and even in the official Challenger Report. The perception that most other human groups were inferior prevailed, even among scientists who prided themselves on rational thought and precise observation, experimentation, and deduction. Before Darwin, the idea that species never changed, that biological determinism controlled all living things, including humans, probably helped reinforce this view. Even after it became clear that species evolve, many European scientists retained their strong belief that the state of human societies, especially their perceived "primitiveness" or lack thereof, had a biological origin.

The *Challenger* scientists' dealings with aboriginal peoples therefore ought to be viewed in the context of their time. Moseley in particular, but the others too, were as curious about societies different from their own as they were about the flora and fauna of remote islands. They were probing the unknown, or the poorly known. Their encounters with indigenous peoples were brief, and they observed and recorded as much information as they could about groups they considered exotic. Still, nearly everything was measured against the standard of their own experience of European society: art, craftsmanship, dress, beliefs, food, government.

The Eurocentric perspective on human life, however, was about to undergo a radical shift. In 1883, a few years after the *Challenger* expedition, Franz Boas, a young German physicist who had recently completed his Ph.D. research on how the reflection, refraction, and polarization of light affected the color of seawater, joined a German voyage to the Arctic. He hoped to continue his investigations by examining any differences that might arise under Arctic conditions. But after he was dropped off at an Inuit settlement on Baffin Island in northern Canada, his attention shifted to how the indigenous people of the region perceived the apparently ever-changing color of the sea. In the European literature Boaz had read, the natives of Arctic Canada, too, were often referred to as "savages": they ate raw meat and fish, dressed in animal skins, and lived in snow houses. Living among them for the better part of a year completely changed Boaz's thinking. He soon realized that although Inuit society was hugely different from his own, the people were far from the "savages" he had read about. He soon abandoned his physics research and began to focus on ethnography; he was at the beginning of a long and distinguished career as an anthropologist during which he and his students would show that there is as much individual variation within different human groups as there is among them, and that different human cultures—Boaz originated the modern sense in which we understand the word—have been shaped by environment and history, not biology. On average, we are all the same. There is no such thing as race in the sense the *Challenger* scientists used the word.

Challenger was in Australia for almost three months, first in Melbourne and later in Sydney. Some dredging was done along the coast, but the naturalists and ship's officers also made excursions inland both for work and for pleasure. It is at this point in his journal that Moseley began to devote much of his attention to anthropology.

He had earlier written briefly about the kitchen middens he saw along the coast near Cape Town, and the stone tools he found scattered through them; he believed the middens were the remnants of ancient Bushmen settlements. In Australia he found even more to pique his interest. On his first excursion into the bush outside Melbourne he commented on the "curious sight" of lines of notches running up the trunks of eucalyptus trees. They reached astounding heights, and they were, he learned, the "tracks" of aboriginals, who carved them as an aid for ascending the trees in search of opossums or honey. Many were old, incised into the trunks of now dead trees, and they made him think of the footprints of extinct animals. His journal entry describing these features is a good example of how deeply the prevailing belief that European society was superior to others was embedded in his thinking: he cannot help comparing the agility of the native people with that of Australians of European descent. Moseley's bias seems to be an entirely unconscious response. He marvels at the climbing ability that must be required to scale the tall trees with so little foothold, but goes on to say, "I was assured . . . that there was a White man in the neighbourhood who could beat any Black at this sort of climbing." He makes a similar comment about boomerang throwing, and concludes: "In fact, a White man, when he brings his superior faculties to bear on the matter, can always beat a savage in his own field, except perhaps at tracking." Moseley and his colleagues were so steeped in the existing European worldview that they were blind to other possibilities. Interestingly, Campbell seems to have been a little more open-minded. When the ship visited Tonga he was impressed by the native men who came alongside in their boats. They were, he observed, "a novel and splendid picture of the *genus homo*; and as far as physique and appearance goes they gave one certainly an immediate impression of being a superior race to ours."

Just north of Sydney is a large estuary into which flows the Hawkesbury River. Numerous smaller streams also empty into it, and the shoreline is intricately scalloped with bays and inlets. Moseley spent several days on one of the side streams, Browera Creek. He was particularly interested in the mix of marine, freshwater, and terrestrial plants and animals there; although it was twenty miles or more from the sea proper, he found it impossible to tell where the creek ended and the estuary began, where the water was salty and where it was fresh. You could sit in an overhanging eucalyptus tree, he wrote, thinking yourself far inland, and fish for sharks. It was an extraordinary place. He wondered what some future geologist might make of the confusing collection of botanical and biological debris that he imagined was accumulating in the sediments below: there would be terrestrial animals and birds, insects, leaves and branches of land plants, freshwater fish—but also saltwater fish and marine shells. As a fossilized assemblage this confusing agglomeration of remains would be difficult to interpret. But something else he had heard about drew Moseley to Browera Creek. In the distant past, over long periods of time, the abundant oysters, mussels, and fish had attracted native people. The area around the creek abounded with large mounds of shells, kitchen middens similar to those he had seen in South Africa. Some of them had been "mined" by the colonists, who sorted out the calcium carbonate shells and burned them to make lime. But many of the shell middens had remained more or less intact, and Moseley sifted through them and collected several stone hatchets to add to his anthropological collection. He also investigated numerous caves that had formed along the streambank by the weathering away of soft sandstone layers. In and around the caves was campfire debris, and when he dug into it he found more relics, including a human thigh bone—something else he kept for the *Challenger* collection. He described and sketched figures that had been drawn on the walls of the caves—fish, animals, even a human form with a tall hat, possibly a European. He wrote of the ubiquitous hand marks, red hand shapes on a white background, formed by spraying a mixture of white clay and water over a hand placed on the red sandstone of the cave wall. Moseley observed that they were the subject of much discussion, but no one seemed to be able to explain why they were there.

Evidently such experiences whetted his appetite for further anthropological investigations. He was disappointed that his only direct contact with aboriginals in Australia came during a visit to a government reserve, where he watched a group of native people play cricket. They were clothed in European dress, and he found the sight incongruous. In his journal he comments that despite the friendship and hospitality of his English countrymen in Australia, "One could not, as a naturalist, help feeling a lurking regret that matters were not still in the same condition as in the days of Captain Cook, and the colonists replaced by the race which they have ousted and destroyed, a race far more interesting and original from an anthropological point of view." His statement betrays a deep curiosity about the exotic and unknown, but it also illustrates again the detached perspective from which he treats an unfamiliar group of people: not much differently from his biological collections, subjects to be examined and characterized.

As the ship moved into the Pacific after leaving Australia, Moseley had numerous opportunities to indulge his anthropological curiosity. His first extended encounter with native people was in Tonga, or, as the crew of *Challenger* universally called the archipelago, the Friendly Islands (the name originated from the friendly reception Cook had received there a century earlier). As soon as the ship anchored, a small boat came alongside, and Moseley began to examine the characteristics of its occupants. These were the same men who impressed Campbell as being superior examples of humans, and Moseley studiously made notes about their coloration, their size, and especially the details of their heads: lips, eyes, forehead, hair. He was impressed by the "unusually marked development of facial expression" they showed as they talked to one another. He studied the way they used their eyebrows, their foreheads, and their faces generally to express emotions, contrasting their behavior with that of Europeans. An obsessive collector of artifacts, he wanted to obtain a lock of hair from one of the men for scientific study. It was not easy; at first the man he chose suspected that he wanted to use it for some kind of witchcraft. But eventually he was successful. In his journal he writes that the reluctance of the Tonga Islanders was a precursor to similar complications he later experienced when he tried to collect hair samples: "I often had amusing difficulties to contend with, and I suspect that some of the girls, from whom I got specimens, thought I was desperately in love with them."

In Tonga, *Challenger* anchored at the island of Tongatabu for four days. Moseley collected plant specimens, described the local fauna, and investigated the coral bedrock of the island and how it got transformed into productive soil. But he spent much of his time studying the people, both through direct observation and by querying resident missionaries. He was especially interested in languages and tried to learn a few words and at least the rudiments of language structure in many of the places he visited. He was struck by the expressive nature of Tongan conversation and how Tongans incorporated gesture into their speech. Without gesture, he concluded, their communication was incomplete.

But language was far from the only aspect of Tongan society that caught his attention. In his journal he describes Tongan dwellings: their size, the materials they were made of, how they were constructed and laid out. He discusses the peculiar "pillows" he found in most of the houses he visited, simple constructions consisting of wooden cross-rods supported on four legs; the sleeper rested his or her neck on the rods. Ever inquisitive, Moseley tried out one of the pillows for himself. He found it supremely uncomfortable; it was pain-inducing, he reported, more like an instrument of torture than an aid to sleep. Half an hour was all he could bear. And the torture, he noted, was undertaken solely in the service of vanity—the function of these strange contraptions was to prevent undue disturbance to the Tongans' sometimes elaborate coiffures while they slept. Moseley thought the Tongans' hair, especially the men's, was the "most remarkable feature" of their appearance. Often they wore intricate jumbles of curls sticking straight up from their heads, colored with yellow and red pigment and sandalwood powder. In Fiji, the next stop after Tonga, he found even more exotic hairstyles to write about. "The various methods of dressing the hair are so numerous," he noted, "as to be indescribable"—and then he went on to describe several. One, which included whitening the hair with "coral lime," "reminded one most forcibly of a barrister's wig."

It is obvious that Moseley thoroughly enjoyed examining the culture of the Tongan people. He and others from *Challenger* were often invited into local homes and offered refreshment; he delighted in the islanders' hospitality and their love of practical jokes. He commented that he would have liked to stay longer. For his anthropological collection he bought several stone axes, which were fast disappearing from use because of the introduction of iron by Europeans. Typical of his desire to understand how things worked, he persuaded one of the men to show him how to start a fire using the "friction method." This, like the stone axes, was something from the island's past; by the 1870s the Tongans were lighting their fires with European matches. But some islanders still knew how to create fire in the old way, and Moseley watched and recorded in detail exactly how they did it. The method involved pressing a dry stick against a slab of dry wood and twirling it rapidly. It looked deceptively easy, but like many skills it took practice to perfect, and when Mosely tried it himself the best he could manage was a little smoke; there was no sign of fire. The process, he wrote, was exceedingly hard on his wrists.

Challenger arrived at Fiji a few days after leaving Tonga and stayed for two weeks, spending much of that time surveying and dredging among the islands. Again Moseley went ashore. He conscientiously fulfilled his duties as a naturalist, collecting and describing specimens for the expedition's collections, but in his journal these take a back seat to his observations of native people. The chapter of his journal on Fiji, more than any other, highlights his fascination with "primitive" cultures.

One of the customs he discusses at length is the ubiquitous consumption of the mildly intoxicating drink kava, which was and still is popular throughout the islands of the South Pacific. Moseley had been introduced to kava in Tonga, and in Fiji, whenever he visited a small village he and his traveling companions would invariably be invited into the home of the local chief to share a cup of the beverage.

The invitation was a simple act of hospitality, but the serving of kava was usually immersed in ceremony. A strict hierarchy was always observed. Everyone sat in a semicircle on the floor, the chief at the head and guests from *Challenger* on either side. Relatives and villagers filled in the semicircle, and the "lower orders" and servants sat farthest away—this, at least, was Moseley's assessment of the local hierarchy. The kava, made from the root of a kind of pepper that is native to the South Pacific, was always prepared in the traditional way, which involved first chewing it thoroughly into a pulpy mass and then briefly infusing it in water. In Fiji, in Moseley's experience, the chewing was always done by young men with very good teeth, and they were scrupulous about hygiene, carefully rinsing out their mouths and washing their hands before they began. The chewing itself required great skill. It had to be done in such a way that very little saliva was mixed into the root, so that it came out as a dry, finely ground mass. Usually several chewers participated in these ceremonies. The products of their efforts were combined in a large wooden bowl and mixed with water; after a short infusion any remaining pieces of solid root were strained out by hand, and the resulting murky brown liquid was placed before the chief. Using a coconut shell, he scooped up some of the kava and took the first drink, at which the assembled villagers clapped. Then each of those present was served a portion of kava in turn, starting with the honored guests from *Challenger* and progressing around the circle.

Kava has a bitter aftertaste, and initially Moseley found it a strange and unpleasant drink. But it was so common in the South Pacific islands that he soon became accustomed to it. His journal entries describe many instances in which he partook of kava as a guest of local people, and he quickly learned the proper protocol: one should drink the kava in one swallow and then put down the coconut shell, spin it around, and say (in the Fijian language) a phrase that he took to mean "the cup is empty." The strength of the kava, he noted, varied considerably depending on the plant used and the way it was prepared. But its effects were similar to those of alcohol: "The gait becomes very unsteady," he reported, and "an elation of spirits is produced."

Moseley seemed unperturbed by the traditional method of kava preparation—chewing the root—but he reported that in Tonga the missionaries were so horrified by the practice that they banned it, insisting that if kava was to be drunk at all the root had to be ground, not chewed. Moseley thought the traditional method produced a far better drink; he considered the kava he sampled in Tonga, made from mechanically ground root, inferior to the Fijian product. George Campbell, on the other hand, had a different view of kava making. During his wanderings in Fiji he was several times invited into family homes and offered kava. In his journal he wrote (I imagine with the usual twinkle in his eyes) that he did not mind if the root was chewed by a pretty young woman, but he was less likely to accept the kava if her husband joined in the preparation.

In the twenty-first century, kava is grown widely in the South Pacific, as well as in Hawaii and a few other places. As with many other things, migration and globalization have spread its use and culture far beyond its origins. Because kava consumption is rarely if ever associated with violence, it has been touted as a safe alternative to alcohol, and kava bars—which have existed for some time on South Pacific islands—have now opened in Europe and North America, especially in California and New York. Kava extracts can be found in health food stores. Fiji and the Pacific island nation of Vanuatu, roughly seven hundred miles to the west of Fiji and at the time of the *Challenger* expedition known as the New Hebrides, have become major exporters to meet the demand. Both countries have long had local kava industries, but they now see cultivation of the plant for export as an important driver of their economies. As a result, to ensure high quality, they regulate both production and export. Kava chewing is not encouraged.

A more gruesome practice, but one that nevertheless intrigued Moseley, was cannibalism. Unlike kava drinking, cannibalism has, thankfully, died out, but at the time of the *Challenger* visit it had only recently disappeared—and from his journal accounts it is clear that Moseley wondered whether it *had* entirely vanished. He recounts hearing reports of feasts on human victims in Fiji less than twenty-five years before his visit, and claims he was told on reliable authority that Europeans had joined in with native people on these occasions. Whenever he had the chance, he would ask native Fijians whether they had eaten human flesh. For the most part he questioned people through an interpreter, which introduced a degree of uncertainty into the conversations, but usually the answer was no. One man he queried, however, replied that he had killed four people but "the chiefs" had not allowed him to eat them. The man, who was probably about thirty-five years old, told Moseley that he felt aggrieved by this prohibition.

Respect, I suppose, or protocol, kept Moseley from asking the same question when he met the Fijian king, Thackombau. The king was an older man and a convert to Christianity, and Moseley found him lively and intelligent. The two men had a long discussion about *Challenger's* scientific investigations, a subject in which the king expressed much interest. However, Thackombau had a darker side: he was rumored to have feasted on some two thousand persons during his lifetime. Moseley had no way to verify the number, which may have been highly exaggerated. But near the king's house was a stone slab that had been used to smash the skulls of unfortunate victims. The slab was set into a base of coral rock that was riddled with small holes and depressions. Moseley noted that the holes were still full of human teeth.

Moseley's journal betrays a gruesome fascination with the details of how cannibalism was practiced by the Fijians. In earlier times, he learned, when a village chief received a distinguished visitor, he would be expected to provide a meal of human flesh. If there happened to be a prisoner handy, he (or she) offered an easy meal. If not, there was another option. Most chiefs would have in their employ a man retained for just such occasions, and he would be sent out to find a victim, usually female, from the surrounding region. Without emotion or any indication of distaste, Moseley recorded that the Fijians took great care in preparing and cooking their victims. They preferred flesh from the thigh or upper arm, and they even had a favorite side dish, a kind of vegetable, that they traditionally served to complement the human flesh. Local young women made the best meal; Europeans were not very flavorful. Moseley's treatment of Fijian cannibalism is as matter-of-fact as that of most things he writes about, whether he is dealing with native cultures or describing a new plant variety. For the most part he is not judgmental; rather, his descriptions and analyses reflect his curiosity and eagerness to learn as much as possible about whatever he is observing. A few paragraphs after he describes the way the Fijians prepare human flesh he writes about a peculiar type of native drum and the bird species he spotted along the course of a river.

When he learned that a huge annual dance festival was to take place on one of the islands while *Challenger* was in Fiji, Moseley was keen to observe, and with help from King Thackombau he and several of his colleagues arranged to attend. The festival had its origins in a traditional yearly gathering to pay tribute to the island's chiefs, but after the arrival of Europeans it was appropriated by the local missionaries, the Wesleyan Missionary Society. The missionaries now used the occasion to celebrate their own "collection day," and the tributes went to them instead of to the chiefs. But in spite of the repurposing, most of the festival's original features had been retained, and Moseley was entranced by the spectacle. He thought he could discern in the performances elements of what he considered to be the more highly cultivated art he was accustomed to in Europe, albeit in a less "developed" form.

When they arrived at the village where the festival was to take place, Moseley and his companions, as honored guests, were invited into the chief's house and offered kava and other refreshments. The missionaries evidently took some care to pay at least lip service to the practice of honoring the chief, because the chief's house was the center of the day's activities, and the dancing took place in a cleared area immediately in front of it. In addition, the chief's young son was given an important role in the performances. As he drank kava, Moseley watched the son being made ready: he was polished with

coconut oil, his hair was trimmed in a Mohawk-like cut that Moseley thought looked like the crest of a Greek helmet, and the rest of
his head was painted a bright red.

When the dancing started, Moseley took notes, recording the performers' movements, chants, and body decorations. The performance included solo and chorus singing, call-and-response singing
that reminded him of a Greek chorus, simple percussion instruments
to accompany the chanting and movement, speeches, and reenact-
ments of past battles, sometimes with real weapons. Some of the
dancers were elaborately painted in contrasting red and black or cos-
tumed in leaves or strips of tappa, the popular Polynesian cloth
made from the bark of the paper mulberry tree. Moseley was sure
that within the elaborate presentation he could identify elements of
multiple art forms that he considered distinct, and he wrote that the
performance contained "the germs of the drama, of vocal and instrumental music, and of poetry, in almost their most primitive condition in development. In these Fijian dances they are all still intimately
connected together, and are seen to arise directly out of one another,
having not as yet reached the stage of separation." He admits that he
is not the first to have pondered the beginnings of the various art
forms: "It was of course not necessary to have recourse to Fiji in order to trace the origin of dancing, music, and the drama. . . . But nowhere, I believe, is the primitive combination of these arts so
forcibly brought before the view, as a matter of present-day occurrence, as in this group of islands." His perception that the "combination of these arts" represented a "primitive" feature is striking;
today multidisciplinary art is common.

The tribute payment was a prominent part of the day's festivities.
Each group of performers represented a village or district, and before starting their dance they would first bring their contributions
to a table set up for the purpose, throwing down coins with as much
clatter as possible. Bystanders shouted encouragement; giving, and
being seen to give, was clearly important. Tribute bearers were subject to peer pressure, especially from local trainees who had been
coached by the Wesleyans. Throughout the day the visitors from

Challenger were pestered by people trying to sell them things—an ornament, a club, a hen—with the goal of obtaining more money to contribute.

As Moseley watched the ferocious looking dances, the performers brightly painted and almost naked, brandishing clubs and spears and uttering piercing war cries, he could hardly believe that this was in fact a missionary meeting. He was amazed at the sway held over the natives by the local missionary, a small, somewhat shabbily dressed and inconsequential-looking European. This man had organized the event and presided over it; he had personally previewed and approved each of the dances that the performers had spent months preparing and perfecting. On the day, the several thousand people who attended the festival were exceptionally well behaved. Moseley concluded that the Wesleyans—who, in addition to organizing festivals, ran a school—were doing good work for the people of Fiji; they should be left in peace to continue with their task.

George Campbell was somewhat less sympathetic. He respected the missionaries, but he was disturbed by the enforced collection he witnessed at the festival. Elsewhere in Fiji he had watched as physical force was used to corral native people into church services. He did not want to "burn his fingers" by writing negative things about missionaries, but these practices did "tell against them in the minds of strangers visiting their fields." Another naval officer, Herbert Swire, who also kept a journal on the voyage, was more overtly critical. "Though I have had but little experience with missionaries, I have already learned, from books and from the statements of officers who have encountered them in all parts of the world, to distrust them utterly; and till *gentlemen* have complete control of their proceedings I am certain they will only cause more harm to the people they are sent amongst than will be covered by the undoubted good which they also bring to them; and this, I believe, is the almost universal opinion in her Majesty's Navy." Swire recorded these thoughts in July 1874 after encountering missionaries at Tonga. He was then in his early twenties; his memoirs were not published until much later, in 1937, but even in that year I suspect his comments raised some eyebrows.

At the time of *Challenger's* visit, rumors were flying that the British were preparing to annex the islands of Fiji. Most of the small community of European settlers who already lived and did business there were in favor of the move, but Moseley noted that the Wesleyans were opposed. They wanted the islands left as they were, and they prayed for that outcome. Moseley did not say whether they supplemented their prayers with more active participation in the debate, but regardless, the prayers were unsuccessful. By the time *Challenger* arrived home in the spring of 1876, Fiji had become British. The king, Thackombau, was taken to Australia on a celebratory trip, and along with the gifts he and his company carried back to Fiji they also brought measles. About a third of the native population succumbed. It was an all too familiar story of colonialism: a one-way transfer of infectious disease from Europe or a European colony to a remote population with no natural immunity.

Perhaps because they were among the first he met, the native people of Fiji, and to some extent also those of Tonga, became benchmarks for Moseley: for the rest of the voyage he would compare each new group of aboriginals he encountered with them. During much of the time the ship was in the Pacific he devotes little space in his journal to the oceanographic work done between island visits, but he writes at length about the language, customs, and physical appearance of the people at each stop.

One such stop was not at an island but at Cape York on the northernmost tip of Australia. Campbell was not impressed with the small settlement there; it was a far cry from Melbourne or Sydney: "Once more in Australia! and a horrid country it would be, if it were all like Cape York." The heat was oppressive, the village ramshackle and sparsely populated. Moseley deemed the natives in a nearby aboriginal encampment to be in a "lower condition" than any others encountered on the expedition. They were listless, wore little clothing, had few belongings, and subsisted on shellfish, snails, grubs, and a local variety of wild bean. But the ever-curious Moseley could always find a few novelties to investigate and write about. One that sparked his interest at Cape York was a peculiar type of bamboo pipe for smoking tobacco, a prized possession for anyone who had one. He made sketches of the pipes, which were large—some were as much as two feet long and three inches in diameter—and required two people to operate. The first smoker lit the tobacco in the bowl of the pipe and, completely enclosing the bowl full of burning leaves in his mouth, blew smoke back into the pipe. When it was full he (or she) handed the pipe to the other, who then inhaled a foot or two of smoke. Moseley wondered at the origin of the strange implement; he had heard reports of similar smoking behavior from elsewhere in Indonesia, and, given the state of the Cape York people, he guessed that the pipes—or at least their design—had been imported. He was also interested in the language of the Cape York inhabitants, which was called Gudang. As an experiment he asked a few of the Gudang-speaking people to count a group of objects. He discovered that they had individual words for *one*, *two*, and *three*, but used the same word for *four* and any higher number. Apparently the Gudang language did not include the concept of a continuous set of numbers; if Moseley gave his subjects more than three stones to count, they would arrange them in groups of two or three.

A little more than a week after leaving Cape York, *Challenger* reached the Aru Islands, in eastern Indonesia. A primary goal at Aru was to collect birds of paradise, but Moseley also found time to indulge his curiosity about language. Here the native people spoke Malay, a language that one of the naval officers, Thomas Tizard, knew from his earlier work in region. He coached Moseley and a few others in the basics, and Moseley proclaimed that Malay was an easy language to learn because it was devoid of grammar. Speakers had no need to worry about niceties such as tense; past, present, and future were one. All a Malay speaker needed to be able to communicate was a dictionary. It was a revelation: "One is irritated on discovering how thoroughly satisfactory such a simple arrangement is, to reflect on the endless complications of verbs and their inflexions in so many other languages and on the time which one has wasted over them." Time was precious to Moseley. There were better ways to spend it than learning verb conjugations.

When the ship reached the Philippines Moseley turned his attention to the architecture of native houses, which were built on stilts; he referred to them as pile dwellings. Influenced by his understanding of biological evolution, he traced out how he thought these ingenious buildings might have developed and evolved over time, from simple safe havens on the water to more elaborate land-based houses on multiple levels. He pointed out that similar buildings could be found around the world, and somewhat bizarrely compared them to Swiss chalets. He hypothesized that their origins went back to a time when native people fled from their enemies by taking to the water in a boat or on a raft. If they had to stay on the water for some time, he supposed, they might anchor in a relatively shallow place using poles, in the same way he had seen Fijians anchor their canoes. From there it would be a short step to building a platform supported by the poles, and the idea of a more permanent structure would be born. The variety of pile dwellings that Moseley saw in the Philippines seemed to buttress his ideas about how they had evolved. Some were set entirely in the water and had to be reached by boat; some hugged the shore and were accessible from land at low tide but water bound at high tide, and some were built entirely on land though still constructed on stilts. The land-based houses had an elevated platform exactly like the water-based ones, and had to be reached by ladder. Living quarters were on the upper level; the spaces below the platform, between the pilings, were often walled in and used for storage or animals (this is the feature that reminded Moseley of Swiss chalets, in which the ground-level space, enclosed in stone walls, was often used for cattle, while the dwelling space was on the floor above). In the more evolved dwellings the wooden piles and some other parts of the houses were constructed of stone or masonry, and the platform became a verandah. But the concept of a house built on piles—or, as Moseley said, the descendants of piles—remained. He watched one being constructed and was struck by the unusual building sequence: for a two-story house the piles were erected and the upper level and roof completed first, the lower level later. It was the reverse of what would

Pile dwellings in the Philippines. (From a sepia drawing by J. J. Wild, courtesy of the Centre for Research Collections, University of Edinburgh.)

be done in England. But the order of construction recapitulated the history of the pile dwelling's development, "just as is the case familiarly in natural history."

By late February 1875, *Challenger* had spent time at Hong Kong, had made two visits to the Philippines, and was about to anchor at Humboldt Bay, New Guinea, before sailing north to Japan. With the exception of the anxious moments experienced during storms among the Antarctic icebergs, this stop in New Guinea was the first place the *Challenger* scientists felt endangered. The ship arrived in the evening, and before long several canoes full of indigenous people appeared. They could not be enticed onto the ship, but they engaged the crew in bartering from their canoes, with various items passed up and down on the prongs of long spears. In the morning, many more canoes appeared in the bay—Moseley, precise as ever, counted sixty-seven—and the bartering intensified. But then a small party from the ship, including Moseley, John Buchanan, and Rudolf von Willemoes-Suhm, headed for shore in one of *Challenger*'s small boats, intending to investigate the local flora and fauna. Every member carried a rifle. As they neared the landing place, a canoe approached, and one of the men in it, standing up and holding a yam, made signs that he wanted to trade. Moseley and the others did not want the yam and signed back no. At this the man acted out an elaborate and, given the circumstances, sinister pantomime. He picked up an arrow, pressed the point against his neck, and then, rolling up his eyes and throwing back his arms, mimicked an agonizing death. Recovering, he again picked up the yam. Trade, he seemed to be saying, or else. Again the men from *Challenger* signaled no. Again the man went through his pantomime. The negotiation was becoming tiresome for the scientists, who were eager to get to work on land. Ignoring the drama, the boat from *Challenger* began to move cautiously toward shore. At this the man in the canoe grabbed his bow, notched an arrow, and drew it back to its full extent. Moseley was standing upright in his boat, and the arrow was pointed directly at him. At close range he was an easy target.

It was a tense moment. In seconds he could be killed or seriously injured. Moseley was holding a rifle, but he did not want to shoot the man, and firing into the air to frighten him was not a good option either: the man might release the arrow. In spite of the danger, though, in some part of his brain Moseley calmly processed and stored his observations. Later he would write them up in his journal. The man, he said, "contorted his face into the most hideous expression of rage, with his teeth clenched and exposed, and his eyes starting." But most likely these gestures were for show, a tactic he used to frighten opponents: "an habitual part of the fight." He may not have been in a rage at all. Other cultures used similar strategies, Moseley noted. In both Chinese and Japanese depictions of battles warriors were often shown with similarly grotesquely contorted faces. It was the same principle of menace that caused wild animals to bare their teeth when confronted.

Even if the man had not been enraged, though, the scientists were still in a dangerous situation, surrounded by canoes full of natives who appeared to be anything but friendly. When one of them snatched a metal box belonging to Moseley, the men from *Challenger* took advantage of the ensuing scuffle for possession of this prize to slip away back to the ship. A lengthy debate followed about the best way to deal with the kind of situation they had just experienced. Moseley was in no doubt that aggression on their part would have resulted in injury, or worse. The arrows of the Humboldt Bay natives were five feet long. For distance shooting they were clumsy, but at short range they were very effective. In spite of the danger, though, a small group from *Challenger* made a second attempt to land later in the day. John Murray was among them, and when they made it to shore he managed to collect several bird specimens. The hostility of the natives, however, limited their stay. The sum total of *Challenger*'s collection from New Guinea was meager, and Moseley's curiosity about the native people had to be satisfied with an examination of the various items he acquired through barter, which included a few bows and arrows and several kinds of stone implement.

At the Admiralty Islands, which lie to the north of Humboldt Bay and are now part of Papua New Guinea, he had better luck. The ship stayed for a week, and because the botany of these islands had not previously been studied, much of his time was occupied with collecting plants. Even so he spent every moment of his spare time investigating the language and habits of the native people. Here he encountered yet another new language; he quizzed people about the words they used and made copious notes. As he had done at Cape York, he ran counting experiments and found that the Admiralty Islanders had a more complex understanding of numbers. Later he published a short research paper on the Admiralty language.

In addition to language studies, in his "spare time" Moseley examined the physical characteristics of the people. He took hair samples and measured the diameters of curls, comparing them with curls of other native people. He measured their height and weight and commented on variations in skin color. He recorded the sizes and shapes of their noses, ears, and teeth. He noted that the men were often much decorated and wore ornaments fashioned from natural materials such as shell or tusk or crocodile teeth. Such adornments echoed the situation in nature, in which male animals were typically the most gaudily colored and decorated. But he could not help poking a little fun. Males in "more highly civilised communities," he commented, were for the most part not ornamented. They "revert[ed] to the savage condition of profuse decoration only as warriors or officials, and on State occasions."

Moseley's investigations in the Admiralty Islands were his last in-depth study of native people on the voyage. The ship would later visit Hawaii and Tahiti, but in both places his contact with indigenous people was limited. He was much interested in Hawaiian myth and legend, and lamented that most representations of the islands' gods had been destroyed by missionaries. He believed that it would have been better if the objects had been sent to European museums for preservation. Still, drawing on what he learned directly in Hawaii and also from his literature research, Moseley was able to speculate about the ways the religious figures—usually carved in wood, or made of wicker—might have evolved over time. He was especially intrigued by the huge mouths on some of these figures, which he thought might have grown larger as religion developed and people made increasingly generous food offerings. He was particularly conscious of the power of change over time. Discussing a strange, nondescript, hook-shaped ornament worn by Hawaiian native people, he speculated that its form could be the result of gradual loss of detail as some original design—possibly even a human face—was successively copied and simplified. Somewhat strangely, but indicative of his constant search for analogies, he claimed that a similar process could be observed on the spires of the Bodleian Library at Oxford: at the bottom of the pinnacles the decorations consist of clearly defined, animal-like gargoyles; these progressively become less distinct until at the top the ornaments are more or less featureless, with no hint of their origin. It is not a very convincing argument, but it is characteristic of Moseley's eclectic, lateral thinking. In a more general sense, considering that his primary responsibilities on the expedition were centered on zoology and botany, his extensive work on the cultures of indigenous people is indicative of his intense curiosity about every aspect of the

world as he encountered it, whether or not it was related to his training or background.

I should say here that Moseley's curiosity about different races was
not confined solely to indigenous aboriginal people. They were more
exotic and less well known to Europeans than the Chinese and Japanese, but when *Challenger* docked at Hong Kong, and later in Japan,
there too he spent much of his time studying local people, customs,
and history. In these places civilization was "more advanced" but no
less interesting. Writing about Japan, he observed, "Notwithstanding
all that has been written . . . the country and its people still remain
almost as great a source of interest and as good a field for investigation as does European civilisation." And he went farther: "The
English and German Asiatic Societies at Japan . . . have still probably
the most fascinating field of research in the world before them."

Challenger arrived at Hong Kong in November 1874 and remained
there for seven weeks. The island had been under British rule since
1841, so it had a strong British presence and a stratum of society that
was familiar to *Challenger*'s officers and scientists. Moseley leaned
heavily on English contacts to help him explore the region and examine local customs. He was as eclectic in his interests in China and
Japan as he had been in the remote Pacific Islands, writing in his journal about everything from language to food, books, and religion.
His constant reference point in these "more civilized" regions was
the habits and customs of his native England, and to some extent his
own experiences. He comments, for example, that foreigners often
find Chinese food "especially nasty," and singles out the famous
century eggs that are pickled and buried for long periods before eating (although in most cases for months rather than centuries). But
the Chinese, he says, would probably be sickened by the meat and
cheese "in a state of decomposition" regularly eaten by the English.
Food preferences were matters of taste and prejudice, he argues; the
English would simply say that the cheese was ripe.

As a biologist Moseley was especially interested in two aspects of
the culture that were quintessentially Chinese and sometimes

intersected: mythical animals and medicine. A friend translated for him ancient Chinese texts dealing with the use of dragons and dragon parts, which, the texts noted, were effective remedies for a variety of problems, from warding off ghosts to curing children of fainting. Moseley traveled north from Hong Kong to visit Guang-zhou, where he searched out druggists' shops selling dragons' teeth and dragons' bones; when he examined them he concluded that they were simply the fossil remnants of mammoths and rhinoceroses. This confirmed his ideas about how mythical animals come to be imagined: they have some basis in fact—such as the fossil bones—but the facts are much embellished through repeated word-of-mouth re-telling. Eventually someone decides to make a sketch of the animal, again embellishing it or exaggerating some feature, and a dragon or a unicorn or a sea serpent is born.

Moseley professed to be one of the few naturalists to have actually seen a sea serpent. He was on a boat traveling from England to Hol-land when a passenger suddenly shouted with excitement that he saw a sea serpent. The passengers crowded to the side of the ship; in the distance they could see a long black object undulating rapidly through the waves. Moseley, skeptical, sought for other explanations. He soon realized that what they were seeing was a flock of cormorants flying close to the sea surface. Waves on the horizon alternately blocked parts of the flock from view, making it look—if one were willing to believe—like a gigantic sea serpent. Chinese literature contains tales of many "most amusing" mythical creatures that Moseley considered similarly illusory. But as the episode of the sea serpent showed, a sometimes unshakeable belief in mythical animals could be found among the English, too. "We are probably not far in advance of the Chinese in this matter," Moseley concluded.

He was not sure the British were much farther ahead in medicine, either. He believed that traditional Chinese medicines containing things like dragon bones were ineffective because they included "a vast number of ingredients, most of them inactive." But he acknowl-edged that it had not been so long ago that ingredients like centi-pedes, lizards, grated human skull, and "all kinds of nastiness" were widely prescribed as medicine in England. Even in his day, "Herbalists still practise upon the uneducated in London, in a style in some respects not very different from that of the Chinese physician."

In Japan, as in China, Moseley wrote about almost everything he saw: religious pilgrimages, the shops selling trinkets at sacred sites, the painted faces of women, exquisite tattoo art, the construction and layout of Japanese houses, details of Japanese theater. Once, though, the tables were turned and Moseley himself became the object of curiosity. On an inland journey, far from any port where residents had become used to seeing foreigners, he was a novelty. With his red beard, he was "especially worth seeing." Some Japanese laughed at the sight, others fled. Mothers grabbed their children and brought them to see the strange phenomenon. Moseley took the interest in good humor. The Japanese tended to be quite uniform in hair and skin color, he said, so it was natural that they would be astonished at "so mongrel a race as the English," with their varied coloring. What is clear, though, is that he was captivated by the scenery, the people, and their customs. He claimed that he had never met anyone, regardless of background or profession, who, having once visited Japan, did not want to return. He implied that he counted himself among such visitors.

The ship remained in Japan for just over two months, visiting several of the country's ports and giving many on the ship an opportunity to travel and learn something of Japanese culture. Campbell came to love the endless cups of tea and even the tiny Japanese pipes, in spite of the fact that they gave him only "one or two whiffs" of tobacco smoke. He was not as enamored of the music. One evening he and his colleagues were entertained by geishas at a hotel; at some point the women took out stringed instruments that Campbell described as guitars. The sound as they began to tune up was, to his ears, truly awful. By comparison, he wrote, "Preliminary bagpipe tuning is the music of the spheres."

Like Campbell, many of the *Challenger* sailors acquired a taste for Japanese tea, and visits to local teahouses became a favorite pastime. (This is one custom Moseley did not describe. If he visited a teahouse, he made no mention of it in his journal.) Joseph Matkin, the steward's assistant, had high praise for one such establishment in Yokohama. It was run by a Miss Midji Maru and her sister, and Matkin was clearly captivated. Miss Maru, he wrote in a letter home, "was very pretty & very genteel: her manner of bowing and receiving payment for refreshment, I have never seen equalled." He was not alone in his admiration. After acquiring her portrait, he learned that at least sixty others from the ship had done the same. Evidently, *Challenger* was good for Miss Maru's business: Matkin wrote that on his last visit to her teahouse—as he paid for his "55th cup of tea"—Miss Maru said "she very sorry when *Challenger* go."

Corals are familiar to most of us. Colorful and beautiful, they are a major attraction for snorkelers and scuba divers on many tropical islands. Aquarium enthusiasts can even grow some varieties at home. Yet although they may seem common, corals occupy only a minuscule fraction of the seafloor, less than 1 percent, and that figure includes vast reefs such as the Great Barrier Reef off the coast of Australia, the world's largest and a place that springs to mind at the very mention of the word *coral*. Even though they inhabit so little of the ocean's physical space, in aggregate the coral reefs of the world are home to around a quarter of all known marine species. They are the rainforests of the oceans.

The naturalists on *Challenger* dredged up their first coral specimen early in the expedition, on January 15, 1873. They had just left Lisbon and were off the coast of Portugal, still in the shakedown phase of the voyage; it was their third dredging attempt. They were too far north to be near a tropical coral reef, and the seafloor was a little over a mile down. When the dredge finally came up, they found it full of "green sand," mostly derived by erosion from the nearby continent. But digging through the mud they were able to pull out a few organisms, including starfish, a species of worm, sponges, and several varieties of coral. One of these was a solitary, deep-water coral of the genus *Flabellum*, and Henry Moseley recognized it as a new species. The convention among biologists is to identify the discoverer of a new species, and this particular species is now listed as "*Flabellum apertum* (Moseley)." It was the first of many new coral species that would be discovered in the course of the expedition; Moseley himself identified seven additional new species just of that one genus, *Flabellum*. Later, in recognition of his work on corals, one of these was named after him: *Flabellum moseleyi*. Fittingly, given his eclectic interests, many species of diverse organisms are named after Moseley, most in recognition of his work on *Challenger*. In addition to the rockhopper penguin and several types of coral, they include a sea cucumber, several varieties of worm, a fly, a fish, and a radiolarian. Those first deep sea corals were a prelude of many more to come. All told, the *Challenger* naturalists dredged up 74 different coral species from the deep ocean bottom and collected representatives of an additional 293 species of shallow-water reef corals. Biologists group corals into various categories, but the division between deep sea and reef-building corals is a fundamental one. Whereas the deep sea varieties can exist over a wide range of depths down to several miles, the reef-building corals—the ones most people are familiar with—can survive only in near-surface, sunlit waters. In clear-water tropical environments this typically means the upper two hundred feet or so of the ocean. Included in *Challenger*'s coral tally are a few species of animals known, somewhat confusingly, as hydrocorals. These are not true corals at all, although they look like corals and are closely related to them biologically. At the time of the expedition biologists were debating whether hydrocorals were genuine corals, and Moseley in particular was interested in conducting investigations into this question.

What, then, are corals? They come in a bewildering range of sizes, shapes, and colors, and inhabit a wide variety of environments, but in the popular conception they are usually thought of as moderate-size objects that might resemble either a rounded boulder or a branching bush. Such structures are communities, however, not individual animals. Within these communities each coral individual shares a simple, basic structure: it is round, it has a mouth and a "stomach"—a simple internal sac for digestion—and around its mouth it has a set of tentacles that aid in feeding. That's it. Waste, anything that cannot be digested, is regurgitated back through the mouth. This structure defines the coral polyp, which looks much like its close relative the common sea anemone. Like the sea anemone, coral polyps draw in their tentacles when they are disturbed, or when they are not feeding. Most polyps are small, but they vary widely in size depending on the variety of coral, from a fraction of an inch up to (rarely) about a foot in diameter.

Many types of coral polyp secrete a hard, durable skeleton of calcium carbonate to protect their soft tissues, and the cumulative mass of billions of these skeletons eventually builds up massive coral reefs, the living corals always occupying the top few feet of a great mound of their dead ancestors. For the *Challenger* naturalists reefs were places of high interest, not only for the corals themselves but also for the complex ecosystems they supported. The scientists' curiosity about coral reefs had in part been stimulated by their countryman Charles Darwin. Darwin is best known for *Origin of Species*, but long before that revolutionary book appeared, and three decades before the *Challenger* expedition, he had published *The Structure and Distribution of Coral Reefs*. This book too contained revolutionary new ideas, arising from his observations when he was employed as a naturalist aboard H.M.S. *Beagle*. One of the explicit objectives of the cruise was to survey atolls, more or less circular coral reefs that are most common in the tropical and subtropical Indian and Pacific Oceans. Why was the British navy especially interested in atolls? These rings of tough, near-sea-level coral reefs typically surround a shallow lagoon and sometimes a low-lying island, and the reefs absorb much of the energy of the crashing waves of the open ocean. They had the potential to provide relatively calm, protected harbors for the navy's ships, far from the nearest land.

Darwin, however, was more interested in the origin of atolls than their usefulness as harbors. Why were they there? How did they get their peculiar structure? Whenever *Beagle* surveyed an atoll, he would examine it in detail from a naturalist's point of view. And he came up with a theory. Atolls, he said, were the final stage of a process that had begun with the growth of reef-building corals around the edges of an oceanic island. Reef corals needed abundant sunlight to survive and could live only in shallow water. If an island began to sink the coral would have to grow upward to ensure that its living parts stayed close to the sunlit sea surface. Conditions would have to be just right—the island would have to sink slowly enough for coral growth to keep pace. If this occurred, a lagoon would eventually develop between the upward-growing reef and the ever smaller

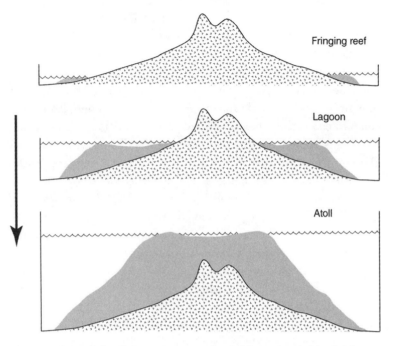

Darwin's theory of atoll formation around a sinking volcanic island. The top panel shows the initial phase of reef growth (gray) around an oceanic volcano (stippled), forming a small fringing reef. As the volcanic island slowly sinks, coral grows upward to stay at sea level. The middle panel shows a more extensive reef, with a lagoon inside a barrier reef. The bottom panel shows a fully developed atoll, the volcano completely submerged. (Modified from a drawing in Charles Darwin, *The Structure and Distribution of Coral Reefs*.)

exposed area of the gradually submerging island. (Darwin's ideal oceanic island was a volcanic cone; think of an inverted ice-cream cone slowly submerging, its above-water area getting smaller and smaller.) Eventually the island would disappear completely and only the reef would remain, a circular atoll surrounding a shallow lagoon, the visible reminder of a vanished volcano.

During his travels on *Beagle*, Darwin saw examples of all stages of this progression, from fringing reefs growing close to the shoreline of an island to fully formed atolls with wide lagoons and no island. His theory seemed to fit perfectly with the observational evidence, and most scientists accepted his ideas. A few years after Darwin's voyage an American geologist, James Dwight Dana, sailed as a naturalist with a U.S. Navy surveying expedition and came to similar conclusions after investigating many coral reefs in the Pacific. Still, there was no hard proof that sunken volcanoes lay below the atolls. Nor could anyone explain why volcanic islands in the middle of the ocean would sink. For these reasons a few scientists remained critical of Darwin's hypothesis, and one of these was John Murray. Murray would develop his own ideas about how coral atolls form.

Scientists now know that Darwin was right. Drilling on atolls inevitably encounters volcanic rock—the vanished volcano—buried deep beneath the coral. And the discovery that the earth's surface is made up of rigid pieces of crust that move around relative to one another, the plates of plate tectonics, has answered the question of why oceanic islands sink. Running through the oceans is a system of long, sinuous, and relatively shallow features known as ocean ridges. The Mid-Atlantic Ridge, curving down the center of the Atlantic Ocean, is a prominent example. All along this ridge system molten lava pours out and freezes into solid rock, creating new seafloor. Although this is a very slow process from a human perspective—about an inch or two of new seafloor is added each year—every part of the ocean floor is constantly in motion, slowly and inexorably traveling away from the ridge where it was created. As it moves away the seafloor cools, and as it gets colder it sinks. The deepest parts of the ocean are the places where the crust is oldest, coldest, and farthest from the ridge where it was formed. Volcanic islands are passengers carried along on the moving seafloor plates, and as the crust sinks, so do the islands. Corals around their flanks grow upward to stay near the sea surface, just as Darwin proposed.

In some places, such as the Great Barrier Reef, a different process has fostered reef growth. From the perspective of the reefs, rising sea level is no different from sinking land; in either case the coral has to grow upward to keep close to the ocean surface. About twenty-five thousand years ago the earth was gripped in its most recent glacial period, and so much water was locked up in the ice sheets that sea level was about four hundred feet lower than it is today. Much of the region that is now the Great Barrier Reef was dry land, part of the Australian continent. But around the continent's edges, especially to the north, the seas were warm enough for reef corals to grow. As the ice age came to an end and the glaciers melted, sea level rose, gradually submerging the low-lying land, and the corals grew upward, keeping pace. The result is the Great Barrier Reef as we know it today. *Challenger*, as she sailed from Vanuatu to Cape York at the northern tip of Australia, had to navigate through these shallow, coral-rich seas. Progress was slow and dangerous; collision with a near-surface reef could rip a hole in the ship's hull. In places reefs had been thrust up above sea level to form low-lying islands of coral sand that were difficult to see in the dark, so the ship had to anchor each night to avoid foundering.

The *Challenger* naturalists understood that reef-building coral species lived only in shallow, near-surface water, but they were unsure about whether there was a depth limit beyond which they could not survive. The consensus was that vigorous growth was probably restricted to about 20 fathoms, or 120 feet. During the expedition a few reef corals were dredged from slightly greater depths, but these were considered anomalies. Why a depth limit should exist, though, was still debated in the 1870s. Some scientists thought that water temperature was a factor, but it could not be the only one because in some tropical seas temperatures were well within the coral-growing window at depths greater than 120 feet. What the *Challenger* scientists did not know was that the reef-building corals have a unique symbiotic relationship with a type of algae known as zooxanthellae, which are photosynthetic and require sunlight to grow and produce nutrients. The corals cannot survive without zooxanthellae, and the zooxanthellae cannot survive without sunlight; this restricts the reef-building corals to shallow water.

The zooxanthellae live within the tissue of the coral polyps, an arrangement that provides them with a protective environment; in return, they produce enough nourishment through photosynthesis to supply the bulk of the coral's, and their own, food needs. This enables the coral polyps to make calcium carbonate skeletons much more quickly than would otherwise be possible—and to build a reef rapidly enough to keep up with sea-level change. To close the circle, some of the waste material from the coral polyps is utilized by the algae. It is an efficient and highly beneficial partnership. It is also a necessary one. Perhaps surprisingly, the clear water of tropical coral reefs, teeming with life, is almost devoid of nutrients. Indeed, it is a bit of a paradox that some of the most biologically productive areas of the ocean have no external source of nourishment. They stand in contrast to the other biological hotspot the *Challenger* scientists observed: the diatom-rich seas of the Antarctic. There the upwelling, nutrient-rich water from deeper regions fuels growth; coral reefs rely on self-sufficiency and recycling. Enormous coral communities such as the Great Barrier Reef could not exist without the photosynthetic algae that use the sun's energy to produce food. Around the coral-algae communities a complex and diverse biological web of interdependent organisms has grown up. Every species plays its part, and the whole exists in delicate balance. Competition for resources is extreme, and small environmental changes have large consequences, which is why scientists are so concerned today about the survival of coral reefs.

But at the time of the *Challenger* expedition, no one worried about the future of the reefs; they were simply interesting natural features that invited close study. And in contrast to the deep corals that had to be retrieved by dredging the distant seafloor, the reef corals could be observed alive, close up, and in context. Moseley seems to have been the one among the naturalists who took the lead. In Bermuda, where some of the northernmost coral reefs grow, aided by the warm waters of the Gulf Stream, he examined corals with the help of a water glass, a small open tube with a glass bottom that he could use from a boat. In the Pacific he waded into shallow lagoons or boated out to a distant reef to study the coral in situ and collect specimens. At the port of Amboina, then part of Dutch Indonesia, he found that the corals he wanted to collect were mostly living at depths of ten or twelve feet, and if he wanted specimens diving down to them was his only option. Afterward his eyes stung for hours. This was long before the era of snorkeling masks, and his face and eyes had been close to the corals as he wrestled them off the bottom. His discomfort, he decided, was probably due to the corals' "nettle-cells." Moseley assumed that fellow naturalists would be among the readers of his published journal, and he provided them with the following advice: "I mention the circumstances [his stinging eyes] as a warning to collectors; where Milleporids are present, great care should certainly be exercised."

Milleporids, in case you are wondering, are a variety of hydrocoral. They are also known as fire corals, the name arising from the burning sensation caused by their sting. Fire corals, jellyfish, and true corals are all members of the zoological phylum Cnidaria (not a tongue twister; the "C" is silent), which comes from the Greek word for "nettle," and all these animals carry Moseley's "nettle-cells." More properly, nettle-cells are called nematocysts, and they, along with many other features of the natural world, are truly amazing biological innovations. Nematocysts consist of small capsules scattered over the tentacles and bodies of coral or jellyfish; if the animal senses danger, or if prey is nearby, the capsules literally explode, shooting out tiny threads toward a target. Different animals have evolved different kinds of nematocysts, depending on their particular needs. Sometimes the threads are barbed and hollow; they embed themselves in the target and deliver a poison. In others the threads are sticky and adhere to the target. Some organisms send out long threads that wrap themselves around the prey, immobilizing it. The poison delivered by a nematocyst can be highly toxic, even to humans, as in the case of the so-called "sea wasp," which is a type of jellyfish, or the Portuguese man-o'-war, which looks like a jellyfish but is not. You do not want to tangle with either of those creatures.

The Challenger Report includes two long sections dealing with corals, one by Moseley, the other by a man named John Quelch, a coral expert at the British Museum. Quelch was given the responsibility of examining and describing *Challenger*'s collection of reef corals—Moseley would deal with the deep sea corals—and, as was the case for some of the other scientists chosen to study parts of the collection, the undertaking was both a prestigious assignment that recognized his expertise and a sometimes burdensome task imposed on an already full research schedule. Quelch was about to leave the British Museum, having recently been appointed curator of a museum in Georgetown, British Guiana (now Guyana). Preparations for the move were stressful, as is apparent from his comments in the preface to his report, in which he apologizes for what he perceives as its incompleteness. His time was limited, and his studies were frequently interrupted, and it was "almost with a feeling of despair" that he managed to get his report into publishable condition. Nevertheless, he took consolation in the realization that despite its shortcomings, "The report will yet form a small contribution to our knowledge of the distribution of the Reef-Corals."

Quelch was not mistaken. He examined every specimen collected during the expedition, determined its biological classification, and either assigned it to an existing species or reported it as new. To make his determinations he worked entirely from the physical characteristics of the coral skeletons, as soft tissues were not available for most specimens. His report is largely descriptive: he measured the skeletons, noted their shape and symmetry, described the arrangement of polyps, and sketched and recorded each unique feature. He traveled to museums in continental Europe to examine their collections and compare specimens with those recovered by *Challenger*. He consulted with other experts in coral biology. Because they were relatively accessible, reef corals had been studied extensively by others prior to the *Challenger* expedition, and the collections made during the voyage were not particularly large. Nevertheless, by the time he was finished Quelch had described seventy-three new species.

One of his most interesting discoveries—from the point of view of someone deeply involved in the study of corals—was an unusual specimen that had been collected in shallow water in the Torres Strait, close to Cape York. It was the only one of its kind found during the expedition, and unlike most of the reef corals Quelch examined, which typically consisted of a large number of very small corallites (the cuplike carbonaceous skeletons of the individual coral polyps that make up a colony), it had only a few large corallites. The whole thing looked like an irregular, wilted disc with a bumpy surface. Quelch knew of no other living coral that was quite like it, but he thought it had affinities to an extinct variety only known from the fossil record. He decided it should be classified as an entirely new genus, which he named after Moseley: *Moseleya latistellata*.

Nearly a century and a half later, the characteristics and distribution of *Moseleya latistellata* are better known. Its range is limited, restricted to the region stretching from northern Australia to Malaysia, Vietnam, and the South China Sea, but even within that area it is rare. The International Union for the Conservation of Nature ranks it as a vulnerable species, the next step down from endangered. Because it lives over a limited depth range—from the surface down to about thirty feet—and is often not firmly attached to the bottom, it is particularly susceptible to the severe storms and warming seas caused by climate change. Whether *Moseleya latistellata* will survive even another half century, to the two hundredth anniversary of the *Challenger* expedition, is by no means certain.

Moseley was not the only *Challenger* naturalist Quelch honored when he chose names for new species. He named a new species of *Millepora* after John Murray, writing that he did so in recognition of the fact that Murray had "materially aided Professor Moseley in his researches on the genus." The collaboration he referred to had occurred when the ship was at Tahiti, where Murray collected live samples from the reef and transferred them to a container filled with seawater for further observation. This was not as simple as it might sound, because these animals are normally so hard to break off the substrate they are fastened to, and so sensitive to even short exposure to air, that they rarely survive the treatment. Moseley had tried to collect living examples several times without success. But Murray succeeded, and his specimens remained alive. Eventually they even extended their tentacles, which had closed up instantly when they were first collected. This gave Moseley, who was interested in the details of their anatomy, an opportunity to study them at close range in their natural state. By the time he managed to examine them, however, *Challenger* had already raised anchor and was under way again, and he had to contend with the inevitable difficulties of conducting sensitive observations on a moving ship. "The animals remained expanded about two hours," Moseley wrote, "but the motion of the ship interfered considerably with the investigation of them." Still, he was able to study them microscopically and gain insights into functions that he could not observe in his studies of dead individuals. Like Quelch, in his section of the Challenger Report he acknowledges Murray's help, writing that he would not have been able to complete his study without it.

Moseley was interested in Milleporids because, as already mentioned, controversy existed about whether they were true corals. This was a question he could address by examining their soft tissues, which would reveal function. For many of the deep sea corals Moseley studied, only the hard calcareous skeletons were available, their soft tissues wholly or partly destroyed by the trauma of their rapid journey from the seafloor to (for them) the very low-pressure environment at the surface. But the soft tissues of the shallow-water Milleporids were almost always complete and available for examination, even if the corals died during collection. Much of Moseley's work on these animals was done during the expedition in *Challenger's* natural history laboratory, and he used all the techniques at his disposal. First he hardened the soft tissues in alcohol or glycerin or some other substance. (One of these substances, according to his notes, was a highly toxic compound of the metal osmium; it is still used today by microscopists as a biological staining agent, but under strict health and safety rules that were not in place in Moseley's day.) Once the tissue had been hardened, Moseley sliced it into thin, transparent slabs for microscopic investigations. By coupling his direct observations of living specimens with high magnification examinations of the cut sections, he was able to work out the structure of *Millepora*'s soft parts in exquisite detail. In his report he describes the different tissue layers, the location and orientation of muscle fibers, the types of tentacles and nematocysts, and the intricate system of tiny "canals" that connect the polyps of the colony. He identified two distinct types of polyps in the animals, one that resembled those of a true coral, with a simple stomach sack and a mouth surrounded by tentacles, and another with no identifiable mouth. The mouthless polyps were ringed with multiple long, thin tentacles. Here was a division of labor: the tentacles of the polyps without mouths were the food gatherers; they ensnared prey and passed it on to the polyps with mouths and stomachs for digestion. This structure was quite different from that of the true corals.

Moseley found other differences too, and they enabled him to answer the question about the status of *Millepora* to his own satisfaction: they were not true corals. One of his regrets, though, was that his soft tissue studies did not reveal the details of *Millepora*'s reproductive organs. Like the true corals, *Millepora* has two modes of reproduction. One is asexual, and in both true corals and Milleporids asexual reproduction is accomplished through a process known as budding, which is essentially cloning; the offspring are identical to the parent. But sexual reproduction in the Milleporids is complex, and unlike that of true corals. Tiny, vase-shaped structures on the Milleporids' polyps release what look like miniature jellyfish into the surrounding seawater. These small structures host the reproductive organs, and they release both eggs and sperm into the ocean; after fertilization the eggs hatch into free-swimming larvae. Following a short but dangerous period of freedom (the sea is full of predators) the larvae fasten themselves to something solid and begin to build a colony.

The true corals do things slightly differently; they release tiny bundles of eggs and sperm directly from the coral polyps. When the release happens as a mass spawning event in a coral reef, it can be spectacular. A film of one such occurrence on the Great Barrier Reef was made for the BBC television series *Blue Planet II*. It documents

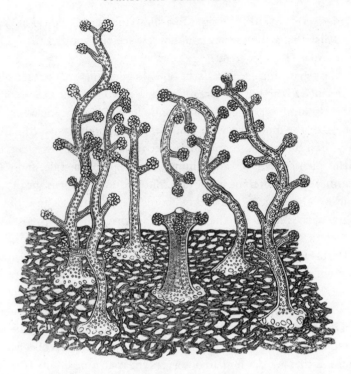

Moseley's magnified, idealized sketch of the surface of a living Milleporid specimen. The short cylindrical polyp in the center has a mouth surrounded by four short tentacles; the longer, sinuous polyps are mouthless and bear many tentacles. These mouthless polyps catch and transfer food to the mouth-bearing polyp. Moseley's histological section showed that the different varieties of polyps are connected at their bases by a series of canals through which nourishment flows to the whole colony. (Courtesy of the Centre for Research Collections, University of Edinburgh.)

one of nature's great spectacles, something that can only be fully appreciated by direct observation, but through the expertise of *Blue Planet*'s camera operators viewers can almost imagine that they are present. The photographers recorded millions upon millions of tiny egg and sperm bundles being released simultaneously from the coral polyps. The bundles filled the water so completely that the camera crew likened the sight to a snowstorm, a biological blizzard so dense that at times they could barely see one another from a few feet away.

Mass spawning events like the one filmed for *Blue Planet* take place only once a year. On the Great Barrier Reef they occur in late spring or early summer, triggered by a confluence of factors, including the phase of the moon: spawning occurs at night, always within a few days of a full moon. Increasing water temperatures as the southern hemisphere warms toward summer, the height of the tide, and water salinity also seem to be important. But whatever governs the precise timing, these events are highly choreographed. Corals in different parts of the reef release their eggs and sperm at different times. Even different species in the same location spawn at different times, presumably to prevent hybridization. The clouds of eggs and sperm rise toward the surface—a snowstorm in reverse—where fertilization takes place, producing tiny larvae that for a period lasting up to several weeks are free-floating organisms. Finally, though, they settle on the bottom and begin the cycle again, cloning themselves by budding to build a new coral colony. But few of the millions of eggs ever make it that far. For human observers the spawning events are spectacular natural phenomena, but for other creatures on the reef they are a rare feast, a time for gorging. The eggs, the sperm, and the larvae are tiny, but they are nourishment. The ones that are not eaten might be washed out to the deep sea or thrown up on a beach in a storm. But numbers count; the millions of eggs and sperm are a kind of insurance policy that nature has devised so that the species will continue: even if most of them do not survive, enough larvae will endure to start new coral colonies.

In addition to investigating *Millepora*, Moseley took charge of examining and reporting on all the deep sea coral species collected during the expedition. Of the seventy-four species dredged up, half had not been described previously. He was especially interested in these organisms because he thought they might help answer the question of whether the deep sea harbors creatures that had been considered

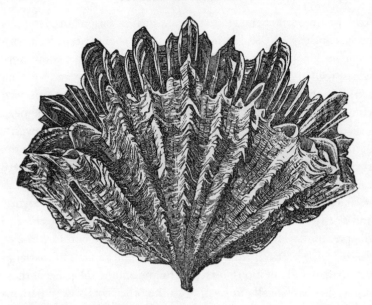

One of the deep corals collected on the expedition and studied by Moseley. He identified it as a new species, *Flabellum alabastrum*, and in his description he says it is "of a beautiful light-pink colour, and is very thin and fragile." It was dredged from about six thousand feet in the Atlantic, near the Azores. (Courtesy of the Centre for Research Collections, University of Edinburgh.)

extinct. But anyone anticipating that "living fossils" would be found in the *Challenger* dredge hauls, corals or otherwise, was disappointed. Among the deep corals, some dredged from depths as great as three miles, Moseley found no "extinct" forms and none with affinities to extinct forms; he also found no correlation between sea depth and *any* index of primitiveness.

Moseley's work on corals is instructive in helping us understand his motivations and also those of the other *Challenger* naturalists. In spite of the substantial existing literature on living corals, Moseley, in the midst of his multifaceted research on things as diverse as botany and anthropology, devoted much time and effort both during and after the expedition to studying these animals. Why did he think this work was important enough to warrant such attention? In part it was because nearly all the existing information about corals dealt with reef corals, not the deep sea ones he was examining. Reef corals were widespread in tropical regions, relatively easy to collect, and had long attracted the interest of biologists from cool European and American climes. Much had been written about them. In contrast, the deep sea corals he focused on were much more poorly known; by investigating them he was breaking new ground. He was experiencing the thrill of finding new species, naming them, and determining where they fit into the overall scheme of coral biology. In light of Darwin's ideas about evolution Moseley was keen to identify "ancestral forms" and clarify possible "lines of descent" from ancient corals to those of the present day. Similarly, in his work on the Milleporids, Moseley was driven by curiosity about the unknown and an urge to answer as yet unanswered questions; he strove to understand the place of these organisms biologically in relation to the other corals. And perhaps another factor was at work for Moseley and his fellow naturalists. As Sir David Attenborough, the British naturalist and presenter of *Blue Planet II* and other spectacular television series about nature for the BBC, once said, "An understanding of the natural world and what's in it is a source of not only a great curiosity but great fulfillment."

John Murray did not have the background in biology and anatomy that Moseley did, and he was not as keenly interested in the classification and fine-scale physiological details of the corals as Moseley was. But he was interested in the larger question of how coral reefs form, and his observations during the voyage convinced him that Darwin's subsidence theory, the idea that reefs around islands grew upward as the islands sank beneath the waves, was wrong. Murray agreed that the reef corals could only grow in shallow water, but he believed that the ocean floor was stable, and he did not think that volcanic islands would subside. Indeed, he found evidence that some had actually risen. He proposed that a reef around an island could only start to grow when enough debris had built up to form a shallow platform or shelf around it. The debris, he suggested, would come both from erosion of the island and from the shells of organisms that lived in the surrounding seas. His thinking was strongly influenced by the fact that the deep sea oozes dredged up by *Challenger* were dominated by the shells of near-surface plankton; similar sediments, he was convinced, would build up around any tropical island. When *Challenger* visited Tahiti he carried out a detailed survey of its reef and noticed that the outer slope was littered with broken-off pieces of dead coral. Over time, he thought, the natural process of coral growth and death would produce large amounts of such debris, and the reef would grow gradually outward. Such a progression would explain the existence of islands surrounded by a broad lagoon and a distant outer reef. In addition, soundings during the *Challenger* expedition had revealed many relatively shallow places that Murray suspected were underwater volcanoes. Here too calcareous debris raining down from near-surface plankton would build up until the depth was shallow enough for reef corals to grow. Murray presented his ideas to the Royal Society of Edinburgh and published a summary in the journal *Nature* in 1880. His hypothesis was sometimes referred to as the talus theory, and many scientists accepted it as a possible alternative to Darwin's ideas. However, Murray's theory was flawed and it was eventually discarded. But his interest in coral reefs, stimulated by his observations during the expedition, had consequences that no one, least of all Murray himself, anticipated. It made him a very rich man.

Murray reportedly declared that the British government recouped the entire cost of the *Challenger* expedition, including the preparation and publication of the Challenger Report, from taxes on the enterprise that brought him his wealth. He was making the point that investment in science, scoffed at by some, not only enhances our understanding of the world but occasionally pays off in a tangible economic sense. One careful analysis of Murray's statement came to the conclusion that although he may have exaggerated slightly he was basically correct, even at the time he made his pronouncement (1913). In the years since, the debt has been repaid many times over.

The source of Murray's riches was a phosphate mine on Christmas Island, a small speck in the Indian Ocean two hundred miles south of Java. *Challenger* did not visit the island, and the connections between the expedition and Murray's involvement in this project are complex. But there *are* connections. When he went to work in the Challenger Office in Edinburgh at the conclusion of the expedition, initially as second-in-command to Wyville Thomson and later, after Thomson became ill, as its director, he was burdened with administrative work. But the science always drove him. Among other things, because of his interest in coral reefs, he asked one of his assistants to draw up a list of all oceanic islands for which at least a modicum of geological information existed, especially information about any coral reefs surrounding them. One of the islands on the list was Christmas Island. It was located in the tropics, but it lacked the lagoon and outer barrier reef common to many tropical islands. From all the available information, Murray concluded that it was mostly made of "coral limestone." No rock samples were available for him to verify this, however, so he asked the hydrographer of the navy, at that time a man named William Wharton, if it would be possible to get some. Wharton was happy to help. Soon afterward, in January,1887, a British ship on its way back from an assignment in the Far East was ordered to divert to the island and collect rocks and coral for Murray. By coincidence, the captain of the ship, John Maclear, had been second-in-command of *Challenger*. He and Murray had spent three and half years at sea together, and they knew one another well.

Christmas Island was uninhabited. It was also rugged, rimmed by a series of steep cliffs, and covered in dense vegetation, and Maclear and his men were unable to explore much of its interior. But they collected pieces of coral near the shore and rocks from the beach, and as soon as the ship returned to England the samples were forwarded to Murray. To his surprise, embedded in one of the fragments of coral, Murray found a piece of pure calcium phosphate. Some of *Challenger*'s deep sea dredge hauls had brought up small crystals and nodules of calcium phosphate, but the sample from Christmas Island was different. It had not formed on the seafloor; Murray deduced that it had originated somewhere on the island, had been washed down into the sea, and had subsequently been overgrown by coral. If he was right there could be a phosphate deposit on the island. Murray again contacted Wharton. Would it be possible to send another ship to Christmas Island? This time could the men explore the interior, especially its higher elevations, and bring back rock samples from known locations, not just cobbles from the beach?

Once again Wharton agreed. And once again the ship diverted to Christmas Island had a *Challenger* connection—its captain, too, a man named Pelham Aldrich, had been an officer on the *Challenger* expedition, and he and Murray were well acquainted. The second visit to the island was better organized and more productive than the first. The flora and fauna of Christmas Island had never been properly documented, so a naturalist from the University of Cambridge was dispatched to join the ship. He and a small party were put ashore for ten days of exploration; they cut paths through the jungle, collected plants and animals, and—the impetus for the visit in the first place—searched for rock samples. Finding suitable outcrops was hard work because of the thick soil layer that had developed in the wet, tropical climate, but they managed to make a small collection of rocks from different parts of the island. When the samples reached Murray in Edinburgh, he found pieces of phosphate among them, confirming his hunch that the island hosted phosphorus-rich rocks.

Why was he so interested in phosphate? In part he wanted to understand how the unusual phosphate rock had formed on this remote tropical island, but perhaps more important were the economic implications. British agriculture had discovered the effectiveness of fertilizers, and phosphate was the most important ingredient. As well, during the *Challenger* expedition Murray had learned firsthand about the benefits of phosphate from an unlikely source: the governor of Bermuda, a man named John Lefroy. Lefroy had pursued a military career, but he also had a scientific background and had managed a magnetic and meteorological observatory in Toronto. He had personally conducted extensive magnetic surveys throughout Canada, and he was a member of the British Royal Society. When *Challenger* docked in Bermuda early in the voyage, Lefroy met with the naturalists and took great interest in their work, helping them with their collections on the island. He also gave them a report on Bermuda soils that highlighted their high phosphate content as a prime reason for their fertility. Wyville Thomson included the report as an appendix in *The Atlantic*, his published journal about the expedition, and undoubtedly it surfaced in Murray's memory when he discovered the phosphate-rich rock on Christmas Island. In the late 1890s, demand for phosphate in Britain was high and rising. If the Christmas Island deposits were extensive, they would be valuable.

Murray faced a problem, however: Christmas Island did not belong to Great Britain. In fact, it was not "owned" by any country. It was remote, devoid of resources, and, until Murray discovered its phosphate deposits, of little interest to anyone. Murray, with his characteristic drive, got in touch with a number of influential friends to find out if it would be feasible for Britain to annex the island. One of those he contacted was the duke of Argyll. It is an indication of how small the intertwined worlds of government, the military, science, and influence were at that time that the duke's son was none other than George Campbell, Murray's *Challenger* shipmate.

With Murray and his influential friends pushing for action, things moved quickly. Less than a year after he examined the first batch of Christmas Island samples and found the phosphate nodule embedded in a piece of coral, the British government had agreed to annex the island. On June 6, 1888, a party from a naval ship landed, conducted a brief ceremony, and claimed the island in the name of Queen Victoria. An hour later they were gone, their proclamation of possession inscribed on a board nailed to a tree. Just over a month later a short notice in the *Times* of London announced the annexation. "The island contains valuable guano," the article reported, "but the anchorage is bad." Murray's interest in the Christmas Island phosphate, though—the notice in the *Times* notwithstanding—had been sparked precisely because it *was not* guano. Guano consists of accumulated bird droppings; it is common on many oceanic islands and is often collected for use as fertilizer. The material from Christmas

Island was pure, hard, phosphate rock. This was something quite different, and much more valuable.

Murray had initially become interested in Christmas Island through his research on coral reefs. Information from the two exploring parties, and the samples they brought back, in addition to revealing the presence of phosphate deposits, confirmed what he had gleaned from earlier reports: the island was mainly constructed of "coral limestone," formed from the calcareous skeletons of reef-building coral. Coral, however, lived near sea level. The limestone on Christmas Island formed cliffs high above the sea. To Murray this was clear evidence against Darwin's theory of reef formation through island subsidence. Christmas Island was a counterexample; it had risen out of the sea, not sunk below it.

Subsequent geological studies have shown that the island is volcanic, formed tens of millions of years ago. It has had a long history of vertical movement, both up and down, and over much of its history, as at present, it has been surrounded by at least narrow coral reefs. At times it has been completely submerged, a condition that accounts for its covering layer of coral limestone, formed in shallow lagoons and actively growing parts of ancient coral reefs when the island was entirely underwater. Furthermore, through the ages Christmas Island has been home to colonies of seabirds that produced large quantities of guano. Tropical rainfall leached phosphorus from the guano and carried it down into the limestone and volcanic rocks beneath; chemical reactions along the way caused it to recrystallize as hard phosphate rock. So ultimately guano was indeed the source of Murray's phosphate, but in an unusual way: through a natural process that purified and refined it.

Only a few phosphate-rich rock samples were brought back by the exploring parties, but they were enough to convince Murray that economically valuable deposits must be embedded in the island's limestone. Characteristically, he charged ahead. Even before the annexation of Christmas Island was official, he had formed an association of friends to fund a further expedition to the island, this one private. By the time the annexation had been publicly announced in the *Times*, his venture was already under way. Murray also applied for a lease that would allow him to exploit any deposits he found. That, however, turned out to be far more difficult to obtain than arranging to have the island annexed. Murray's request for a lease thrust him squarely into the midst of bureaucratic wrangling within the government, and it also brought competing claims to light. He eventually got his lease, but it took years, and the final result was not quite what he wanted.

Murray faced several difficulties. First, he was on the government payroll as director of the Challenger Office. Some officials argued that this disqualified him from being granted a private lease, particularly in a territory he had been instrumental in acquiring for the government. A further difficulty was that he was a scientist, not a businessman. Perhaps he was uninterested in mining the phosphate but only wanted the lease to sell on to the highest bidder and thus profit handsomely, the thinking went. Finally, some sixty years earlier a family from Scotland by the name of Ross had settled and set up plantations on the Cocos and Keeling Islands, remote coral atolls in the Indian Ocean six hundred miles to the west of Christmas Island. The Rosses had been highly successful, and by the late nineteenth century they had built up a regional trading empire. Their ships occasionally called at Christmas Island on journeys to and from Java and Singapore, and they considered it to be part of their own little Indian Ocean kingdom. When they learned it had been annexed, their agents in London argued that the Rosses' history with the island gave them precedence over Murray; now that Christmas Island was British *they* wanted the formal lease. The agents declared that the Rosses had earlier (before Murray) urged the government to annex the island, and they also said that they had in fact established a permanent base there. But no additional evidence supporting either claim was produced, and Murray strongly disputed their assertions. Both parties had supporters in the bureaucracy. The wrangling dragged on, and in the end a compromise was reached: Murray and George Ross, representing the family, were jointly given a ninety-nine-year lease. In return they were to pay the government a royalty of 5 percent on any phosphate or timber exports. The terms were not what Murray had hoped for, because he would now have to deal with a partner. But finally he had his lease and could start making plans.

Even then, things did not go smoothly. The Rosses were reluctant partners; their business was plantations and timber harvesting, not mining and geology, and for a long time they took little interest in phosphate. They knew that guano could be used for fertilizer, but they did not understand that the Christmas Island phosphate was different, much purer and more desirable than mere bird droppings. For a while they thought Murray must be hiding something—that what he was really after was gold—and they sent men to the island to look for it. Of course they were unsuccessful. But for years after the lease was signed, they rebuffed Murray's pleas to set up a joint phosphate mining enterprise. Finally, George Ross traveled to London for further discussions, and on January 14, 1897—six years after the lease came into effect—the Christmas Island Phosphate Company was formed.

Murray was its first chairman, and he was eager to get the mining operation under way, but science was always uppermost in his mind. He had the foresight to realize that large-scale mining might alter the natural environment of the island, and one of his first decisions was to arrange through the Natural History Museum in London for a naturalist to be sent out for a thorough study of the island. The man chosen, Charles Andrews, left England a few months after the company was formed and stayed on Christmas Island for ten months. His investigations resulted in a valuable baseline report on its natural history before mining commenced.

While Andrews was conducting his study, a British engineer, also sent by Murray, began mining operations. He supervised a small force of contracted laborers, but for several years progress was slow. Then, in 1900, Murray himself visited. It was the first time he had been to the island. About to turn sixty, he became the first person to hack his way through the jungle from one side of the island to the other, and due either to his inspiration or simply because of the natural evolution of the mine, phosphate production began to rise rapidly a short time later. By 1907 more than one hundred thousand tons were being exported annually. By 1912 at least a million tons of phosphate had been extracted.

Murray remained chairman of the company until his death in 1914, and he was clearly its driving force. The Ross family gradually reduced their active participation; they were content to collect the substantial income their 50 percent share generated. With a few short interruptions, phosphate mining on the island, which is now an Australian territory, has continued up to the present day, although because of possible environmental damage vigorous debate is currently under way about whether the company should be permitted to extend its operations to new, unmined areas. Regardless, for more than a hundred years Murray's company and phosphate mining have been the distinguishing features of Christmas Island, its primary industry and employer. The island's highest hill is named after Murray, and in recognition of his local importance his portrait has twice appeared on the island's postage stamps.

Murray's discovery and exploitation of phosphate on Christmas Island probably eventually brought the government more than enough revenue to repay the entire cost of the *Challenger* expedition, as he claimed. The wealth he accrued personally from the venture also allowed him to privately support multiple scientific endeavors. Even before mining on Christmas Island had begun, he had, with the help of friends, set up a marine laboratory on the shores of the Firth of Forth, near Edinburgh. It was the first such establishment in Britain, and it soon became a center for teaching and research in the marine sciences. With funds from the Phosphate Company, its continued operation and maintenance was assured. Murray supported much of his own research from personal resources: he organized and completed a major study of Scotland's lochs, still recognized today for its accuracy, and he provided financial support for a 1910 Norwegian seagoing expedition in which he participated at the age of sixty-nine. In honor of an American friend and colleague he funded a medal that was awarded through the National Academy of Sciences in the United States. After his death, his estate continued to fund oceanographic exploration and research. All these activities, in the serendipitous way the world works, can be traced back to Murray's curiosity about how coral reefs form.

In the nineteenth century scientists like Darwin, Dana, and those on *Challenger* were interested in the details of coral anatomy, coral evolution, the classification of coral species, and theories of reef formation. Today's scientists are still curious about these things, but they are also concerned about reef survival. Along with a few other organisms, corals are sometimes viewed as canaries in a coalmine, signaling danger ahead. Reef survival is precarious at the best of times, as is evident from the vast overproduction of sperm and eggs in the mass spawning events: the abundance is necessary to ensure that a few will survive to start new coral growth. But even small environmental excursions from the norm can be damaging, and sometimes lethal, for coral and the delicately balanced ecosystems that flourish in and around reefs.

Water temperature seems to be especially important; most corals are healthy only within a narrow temperature range. This was brought home emphatically in 1997–1998 when one of the most powerful El Niño events on record significantly raised ocean temperatures in many parts of the world. Suddenly a phenomenon that had been known for a long time but had been considered a sporadic and local occurrence—coral bleaching—became a global problem, affecting coral reefs worldwide. Scientists estimate that in the neighborhood of 15 percent of the world's reefs were destroyed in that single event.

What is coral bleaching? As the name implies, it's a phenomenon that causes the corals to lose their color; they turn a ghostly white. This happens when environmental stress causes the coral to expel the symbiotic algae that live in their soft tissue—the zooxanthellae. Because these organisms give coral its color, without them it has a bleached appearance. Expelling the zooxanthellae is a defense mechanism; the algae are sensitive to environmental stresses, and as temperatures rise above normal levels they begin to produce chemicals that are damaging to the coral. When this happens, they are no longer welcome. Reefs can survive short bleaching episodes if the zooxanthellae eventually reestablish themselves in the coral, but because the algae provide the bulk of the coral's nutrient needs, long periods without them can destroy a reef. With average ocean temperatures expected to increase into the indefinite future as the planet warms, bleaching will inevitably become more frequent and more widespread, and will last longer. Reefs may not have time to recover from one bleaching event before they are hit with another. In the geologically recent past, coral reefs adjusted to large temperature swings during the cold-warm cycles of the Pleistocene Ice Age. But in comparison to today's warming, those changes were gradual. The pace of change now may be too rapid for the coral to accommodate.

Adding to the threat, steadily rising carbon dioxide in the atmosphere, the primary cause of warming temperatures, has a further harmful effect. With higher concentrations of the gas in the atmosphere, more dissolves in the ocean, and seawater becomes more acidic. This is a measurable effect; scientists have recorded a steady change in ocean water pH as atmospheric carbon dioxide has risen. Every marine organism that makes its shell or skeleton from calcium carbonate is sensitive to this change; the more acidic the seawater, the more difficulty they have in secreting their skeletons. Field geologists have a simple way to determine whether a rock is limestone (calcium carbonate): they test it with a drop of dilute hydrochloric acid. If the sample is limestone it will fizz and bubble as the calcium carbonate dissolves and releases carbon dioxide. Seawater is not as acidic as the geologist's hydrochloric acid, but the principle is the same: as its acidity increases, seawater will actively attack the calcium carbonate skeletons of corals, dissolving and perhaps eventually destroying them. Even if they are not killed outright, the damaged reefs will be less likely to survive the stronger and more frequent storms predicted under most global warming scenarios.

If these realities seem to paint a grim picture of the future for coral reefs, there are nevertheless a few (small) glimmerings of hope. Many of the world's coral reefs are now protected areas or parts of national parks. Their well-being is monitored, and the scientists and reef managers who work in the parks are seeking ways to make them more resilient. Some coral species are more tolerant of high water temperatures than others, some do better in shaded locations than in direct sunlight. Understanding such differences may help in the artificial repopulation of a bleached area after a warming event. Still, it is a massive task. During 2014–2017 a prolonged period of ocean warming due to another strong El Niño event caused further widespread damage to reefs around the world, perhaps the worst ever recorded. Divers in the northern parts of the Great Barrier Reef said the effects were catastrophic: dead and dying organisms everywhere, the stench nauseating. The Coral Watch program of the U.S. National Ocean and Atmospheric Administration reported that during this period more than 70 percent of the world's coral reefs were subjected to heat stress that could potentially cause bleaching and/or mortality.

The *Challenger* naturalists of a century and a half ago—a mere eyeblink on evolutionary or geological timescales—knew nothing about coral bleaching. It was not a part of their vocabulary. Today, coral bleaching is a familiar and worrying phenomenon to anyone concerned about global warming and the earth's biological diversity. The Challenger Report contains no hint that any of the reefs examined during the expedition were damaged or in poor condition. Today a visitor to almost any reef in the world cannot avoid seeing the effects of bleaching, a startling reminder of how rapidly and extensively our own species is altering the natural environment.

The main focus of *Challenger*'s mission was the deep sea, but in spite of this, and in spite of Wyville Thomson's comments about discouraging work on land if it interfered with that mission, the *Challenger* scientists spent a fair amount of time on shore. Much of it was in South Africa, Australia, Japan, and China while the ship was being provisioned and repaired, but oceanic islands held a special lure for the naturalists, with good reason. Islands, particularly isolated oceanic islands, were—like the deep sea—places where they knew they had a high chance of finding new and interesting species.

Henry Moseley, as we have seen, had a special interest in islands, an interest that went well beyond his fascination with the indigenous people he might encounter on them. In the concluding remarks to *Notes by a Naturalist on the "Challenger"* he spells out at some length why islands attracted him; indeed, the last page of his book is devoted to the importance of studying islands. He begins his remarks by praising the accomplishments of the expedition and the foresight of Thomson and others in organizing this major study of the oceans. Then comes the caveat. The deep sea, he asserts, is likely to remain much the way it is now for a long time into the future. It merits further exploration, but there is no real urgency; it can be investigated at leisure. Its physical features are unchanging, and its plant and animal life might well still be in their present condition "long after man has died out." The situation on dry land, however, especially on islands, is different. There, "animals and plants and races of men are perishing rapidly day by day, and will soon be, like the Dodo, things of the past." (The dodo, a favorite example of a species that had become extinct at the hands of humans, had inhabited the remote Indian Ocean island of Mauritius until it was killed off by sailors and invasive species; the last known dodo was spotted in 1662.) Moseley suggested that an expedition dedicated to exploring and cataloguing life on isolated islands, especially in the Pacific, was an urgent necessity. He noted that even in the Atlantic, which was essentially in Britain's backyard, there was still an island—though he does not name it—whose flora and fauna had never been studied. In his opinion, this situation was "to the disgrace of the British enterprise."

The expedition he proposed would yield valuable knowledge about remote islands, but only if it were launched in the near future, "before introduced weeds and goats destroyed their [the islands'] natural vegetation; dogs, cats, and pigs their animals, and their human inhabitants [were] swept away, or . . . had their individuality merged in the onward press of European enterprise." His apocalyptic vision drew on his experiences during the expedition. He had seen firsthand the effects of introduced domestic animals on several islands, and in the Pacific he had observed the impact of missionaries and traders on the way of life of native people. His fascination with how plant and animal species developed in isolation on islands also led him to another conclusion: "The earth, considered as a comparatively insignificant component particle of the universe, may be justly compared to a small isolated island on its own surface." This is a prescient thought, a sentence that could easily have been written a century later by a scientist reflecting on the iconic image of our small blue planet taken from an *Apollo* spacecraft.

The importance of islands for the biological sciences was highlighted by the work of nineteenth-century world travelers such as Moseley's countrymen Charles Darwin and Alfred Russel Wallace. Darwin's observations from the Galápagos are well known. When he visited the volcanic islands, situated some six hundred miles west of the nearest land, he was astounded by the diversity and strangeness of the life he found there. Many of the plants and animals were new to him. He realized that, geologically speaking, the islands were relatively young, and at some time in the geologically recent past, when the volcanoes had just poked their heads above the surface of the sea, they must have been nothing but barren rock. The diverse life he observed had arrived by accident: seeds blown on the wind or carried by errant birds, themselves swept off course by storms;

reptiles and insects perhaps ferried to the islands on floating debris. Even the rats and bats, the only terrestrial mammals, could plausibly have been transported westward from South America. But the islands had no large land mammals. Such creatures would have had no way to cross the wide sea barrier.

Darwin was also struck by the island-to-island differences he observed in the flora and fauna within the Galápagos archipelago. Contrary to some popular accounts, however, he did not immediately recognize that the variations he saw—for example among the now-famous finches—actually marked different species that had evolved to fill distinct ecological niches. His thinking was still constrained by the prevailing idea that species were immutable, and it was only years later, after he returned to England and he and various expert colleagues had closely examined his collections, that he formulated his ideas about species evolution by natural selection.

At about the same time, Wallace was working on the islands of Malaysia and Indonesia. These were very different from the Galápagos. They were not as isolated, and—although Wallace did not realize this—lower sea level in the past had provided land bridges, corridors for migration, between some of the islands and even to nearby continents. Wallace traveled extensively in the region, making observations and amassing a large collection of specimens. (This was partly from necessity; to help fund his investigations he collected duplicates and sold them to collectors.) In his work he documented island-to-island differences for many organisms, and his observations are often cited as laying the groundwork for a new discipline, biogeography: the science of how and why biological species are distributed in space and time. The variations Wallace observed among and within the islands of Malaysia and Indonesia led him to conclusions about speciation that were similar to those reached by Darwin based on his Galápagos experience. The two men, working in distant parts of the earth and investigating very different assemblages of living things, developed similar interpretations of their observations.

It is no accident that Darwin's and Wallace's seminal ideas grew out of island studies. The process of evolution does not radically differ on these often remote specks of land, but because they are small and isolated from outside influence, unique, clearly observable evolutionary changes have often occurred on them. They are sometimes described as natural laboratories for evolutionary biology. When the *Challenger* expedition began, ideas about evolution and species distribution were in a state of flux and still controversial. The question of how different species came about, and especially of where humans fit in the scheme of living things, had been debated for decades, and the blockbuster, Darwin's *Origin of Species*, had appeared just thirteen years earlier. Still, the journals and reports of the *Challenger* scientists show that for the most part they embraced Darwin's and Wallace's ideas. This is particularly apparent in the case of Moseley, perhaps because his popular book focuses on land plants and animals, especially those of islands, rather than on the flora and fauna of the deep sea. That perspective gave him plenty of opportunity to speculate on evolutionary themes.

It was already well known in Moseley's time that many island species are endemic—that is, their occurrence, as was true for the dodo, is restricted to a single island or a cluster of geographically closely related islands such as the Galápagos. Following Darwin's and Wallace's lead, Moseley was especially curious about how the ancestors of those endemic species first arrived on isolated islands, and how they then developed and evolved. His observations from remote Marion Island, which the ship visited in late December 1873, offer a good example. The island lies at 43° south latitude, on the northern fringe of the Southern Ocean. Moseley writes about the relatively abundant vegetation and notes that some of the species on the island have similarities to land plants of southern South America and the Falkland Islands. But the distance between these places is vast— Marion Island is over four thousand miles from the Falklands, and farther still from mainland South America. If the ancestors of some of the Marion Island plants came from South America, how did they get there? Moseley thought they might have been transported by the Antarctic Drift (the Circumpolar Current) that sweeps through those latitudes from west to east. Or perhaps seeds were brought by seabirds that traveled long distances on the prevailing westerly winds. These were rational possibilities based on his knowledge of the direction and course of the winds and Antarctic Drift, and the long-distance flying ability of some seabirds. Obviously the plants were there; they had arrived somehow. But the details of how were still speculation and would require further investigation.

Moseley was especially curious about the origins of a peculiar plant known as the Kerguelen cabbage, familiar to early sailors as a preventative for scurvy (its leaves are rich in vitamin C). He saw it first on the visit to Marion Island, but he observed it later on Kerguelen, where it was more abundant, and also on Heard Island. Although these islands are widely separated, they are usually lumped together as a single family of sub-Antarctic Southern Ocean islands, not least because of similarities in their vegetation. Like Moseley, George Campbell comments on Kerguelen cabbage in his journal, though he does so from a culinary point of view. It was served as a vegetable in the officers' mess when the ship was at Kerguelen, but he did not care for it. Neither did most of the other officers, so it quickly disappeared from their menu. But "the men" had a different opinion. They enjoyed the vegetable and ate large quantities of it every night at dinner. Their enthusiasm might simply have been because it was fresh; Joseph Matkin wrote in one of his letters that the scientists and officers generally ate almost as well during the expedition as they did at home while the rest of the crew had to make do with less appealing and often stale, moldy, or weevil-filled fare.

Moseley did not mention whether he sampled Kerguelen cabbage at the dinner table. But he did speculate about its origins. The plant is endemic to the isolated islands of the Southern Ocean; it is found nowhere else. How did its ancestors reach this remote region? And even after it had established a foothold on one of the islands, how did it navigate the long stretches of hostile ocean to reach the others? Moseley examined seeds from the plant and concluded that they were not very robust. He thought they would be incapable of surviving a long sea journey, throwing into doubt his idea about transport from

South America on the currents of the Antarctic Drift. For the same reason even inter-island transport by water did not seem feasible: the distance from Marion Island to Kerguelen is fifteen hundred miles. Moseley noted that Darwin had suspected that seeds for at least some of the vegetation on the Kerguelen Islands had been brought from elsewhere on icebergs, but he dismissed this idea. Perhaps, he thought, there had been land bridges between the islands at some time in the past, facilitating dispersal. This was not an original idea; it had been suggested some thirty-five years earlier by J. D. Hooker, the biologist who accompanied James Ross on his voyage in the Antarctic seas. But *Challenger*'s soundings, especially around Kerguelen, seemed to confirm the possibility. They revealed an extensive shallow plateau in the region; Marion, Kerguelen, and the other sub-Antarctic islands, Moseley decided, could be the remaining exposed peaks of a huge tract of formerly dry land that was now mostly below sea level. This possibility would not solve the problem of how the plants initially reached the region, but it would explain their spread among the now isolated islands.

One hundred fifty years later biologists are still probing the question of how and when endemic species like the Kerguelen cabbage reached the remote islands of the Southern Ocean. The question remains important because the islands may hold clues to the history of Southern Hemisphere vegetation, particularly if they were places of refuge for certain plants during the ice ages of the past few million years, when the Antarctic continent has been devoid of vegetation. Today's scientists have a much better understanding of the geological and climate history of the region than existed at the time of the *Challenger* expedition, and they have much better tools for investigating the origins of individual species, including a crucial tool, DNA analysis. They also know with certainty that no former landmass connected the islands. The shallow plateau that the *Challenger* scientists detected around Kerguelen exists, and geological studies have shown that sections of it were above sea level in the distant past, more than twenty million years ago. But the plateau never extended to the

Kerguelen cabbage (*Pringlea antiscorbutica*), photographed on Kerguelen Island. In addition to the clump in the foreground, several other clumps of the plant are visible, their fronds sticking up vertically. (*Report on the Scientific Results of the Voyage of H.M.S. Challenger . . . Narrative; Volume 1, First Part*, plate 16.)

other Southern Ocean islands. Contrary to Moseley's theories, ancestors of Kerguelen cabbage and other endemic species must in fact have endured very long sea journeys.

In spite of today's scientists' better knowledge and better tools, however, much remains unclear. As recently as 2012, a scientific paper on the endemic plant species of the sub-Antarctic islands referred to the Kerguelen cabbage as "enigmatic." No other member of the cabbage family shares its particular physical and ecological characteristics: the shape, size, and arrangement of its leaves, its adaptation to wet soil and cold climates. The authors of the paper collected samples from all the islands visited by *Challenger* (Marion, Kerguelen, Heard), as well as from the fourth island group in the region, the Crozet Archipelago, and they used DNA sequencing to determine the plant's lineage and estimate the time of its arrival. Their results indicate that the ancestor of today's Kerguelen cabbage did indeed come from South America. The DNA analyses also suggest that it arrived relatively recently in geological terms—within the past few million years—and rapidly evolved into the present-day species that is so different from any other living member of its family.

But the question of how that ancestor initially made it to the sub-Antarctic islands remains. Only a few options seem viable. One is direct dispersal from the southern part of South America. Another is that the Kerguelen cabbage ancestor first dispersed from South America to the Antarctic and used the continent as a "stepping-stone" for later travel to the sub-Antarctic islands. Such a scenario would require shorter over-water travel distances, but so far no plant fossils have been found in Antarctica that would support the hypothesis, so it remains speculation. Regardless of how the species arrived, however, long-distance dispersal was still required to spread the plant among the different island groups. The most likely mechanism—either for direct colonization or via Antarctica as a stepping-stone—is transport by seabirds such as the albatross, which is known to travel extremely long distances. The seeds of the Kerguelen cabbage have a mucilaginous coating, which may help them stick to the birds' feathers or feet. Seeds could also have hitchhiked on land birds: in another part of the world—Hawaii—both bats and land birds managed to travel more than two thousand miles from the nearest continent to reach the newly formed volcanic islands.

Charles Darwin was deeply interested in the question of how plants could populate newly formed volcanic islands, his interest stimulated by his observations on *Beagle*. About twenty years before *Challenger* visited Kerguelen he had conducted simple but extensive experiments, putting a variety of seeds in salt water and recording how long they floated, and then testing whether they would germinate. Many sank within a few days, but some floated for months and grew into healthy plants when later put in soil. Once he even floated a dead pigeon in salt water for a month after feeding it on seeds. This sounds gruesome and it quite probably was, but it was a telling experiment. When Darwin retrieved the seeds from the pigeon's crop and planted them, they germinated normally. Maybe seeds could be transported from a continent to a distant island in the gullet of a dead bird. Moseley never mentions Darwin's experiments, but presumably he was aware of them. Even if the exact mechanism by which the ancestor of Kerguelen cabbage reached the Southern Ocean islands is never uncovered, it seems safe to conclude that one way or another, and given sufficient time, plant seeds from distant continents are capable of reaching the most remote oceanic islands.

Of course the Kerguelen cabbage is not the only "enigmatic" plant species native to the sub-Antarctic islands. Another, which was also described by Moseley, is a kind of "cushion plant" that grows in rounded clumps consisting of multiple small plants and looks like . . . a cushion. Cushionlike plant colonies are a relatively common adaptation to cold, harsh environments—they are better able to survive a difficult climate by massing together—so it is not surprising to find them on the islands of the Southern Ocean. The plant, *Lyallia kerguelensis*, was already known to science, and Moseley searched for it on each of the sub-Antarctic islands visited by *Challenger* but found it only on Kerguelen. (Subsequent research has shown that it is endemic to Kerguelen.) Like the Kerguelen cabbage, *Lyallia* has recently been the subject of DNA analysis, but with quite different results. Genetically, its closest relative is a plant that is endemic to New Zealand. The molecular analysis indicates that these two species split off from a common ancestor twenty to twenty-five million years ago, at a time when the Antarctic continent, cooling toward the ice age but not yet ice covered, still hosted a diverse vegetation cover. The scientists involved in this work infer that the common ancestor of the two plants was most likely widespread in Antarctica, and that as the climate continued to cool *Lyallia* eventually found refuge on Kerguelen. It survived, but its Antarctic ancestors died out as ice gradually covered the continent. Now *Lyallia* is limited to Kerguelen, but in spite of its long history there it has never managed to colonize any of the other sub-Antarctic islands, unlike the Kerguelen cabbage.

Kerguelen cabbage and *Lyallia* offer good examples of why studying the flora and fauna of remote islands was so important to Moseley. Because of their isolation both species evolved in ways that differed from those of even their closest relatives. And both illustrate the ability of living organisms to disperse over huge distances, seemingly against all odds, and colonize new environments. Even their dispersal among the sub-Antarctic islands is a fascinating puzzle: in spite of the fact that the ancestral plants of both species made lengthy journeys over water, only one, the Kerguelen cabbage, managed to traverse the shorter inter-island distances necessary to populate multiple Southern Ocean islands. For a curious naturalist, endemic species like these posed questions that could not be ignored.

Of all the islands visited by *Challenger* during her three-and-a-half-year voyage, those of the Southern Ocean were the most remote. They were also the most barren. On Heard Island, Moseley found only five species of flowering plants and four species of moss. Search as he might, he could not find even a single lowly lichen. In his journal he noted that the island had "a miserably poor flora," and this lack, like the mechanisms of plant dispersal, puzzled him. He wrote in his journal that in both the Southern and Northern Hemispheres many other islands lay at similar or even higher latitudes yet were much more diverse. Biologists had already described 119 species of flowering plants from the Falklands, compared to Heard's 5; on Melville Island, north of the Arctic Circle in Canada, they had identified 67. Obviously latitude did not tell the whole story. Isolation and, especially, climate had important roles to play.

In the Pacific most of the islands visited by *Challenger* were tropical or subtropical, and the flora and fauna were abundant. Here Moseley had no complaints about diversity. Many of the species he and the other naturalists encountered there were endemic. But the degree of endemism varies greatly from one island or island group to another. On the Hawaiian Islands over 90 percent of the native flowering plants are endemic; on the small and remote Cook Islands in the southern Pacific the figure is around one percent. How do such differences come about? There is no unanimity among experts, but they do agree about some general principles.

The first and arguably most important factor has to do with the initial arrival of new colonizing species. They might consist of land birds blown off course in a storm, or plant seeds adhering to a seabird, but whatever the mechanism, the arriving immigrant population is likely to be small, and the environment they are about to inhabit will almost certainly be different from the one they left. The more remote the island, the less frequent such accidental arrivals will be. By its nature, dispersal itself is a filter that helps determine what kinds of organisms inhabit islands. A bird, an insect, a plant, or even a small mammal might inadvertently cross a wide stretch of ocean and reach an island, but what about a freshwater fish? Or an elephant? Not so plausible.

In terms of modern understanding, the limited number of new arrivals means that the gene pool they bring with them is small and is also a biased sample of the much greater genetic diversity that exists among the parent population left behind. As the arrivals become established, their limited numbers ensure that normal genetic drift—the random generation-to-generation changes that occur in all living things—drives genetic changes much faster than it would in a large population. Coupled with mutations, which also occur regularly in all organisms, and natural selection as the migrants adapt to new surroundings, large differences can quickly develop between the newcomer population and its ancestors from the mainland. Even if occasional further accidental migrants from the parent population reach the island, the impact will be minor. Over time, the island population will probably evolve into a completely new species. If the island or group of islands has a range of topography and microclimates—dry regions, wet regions, tropical lowlands, cold volcanic mountain peaks—natural selection may drive even further speciation.

Darwin's Galápagos finches offer an especially good example. Research on their genetics indicates that initially a tiny group, perhaps only thirty or so individuals, arrived on one of the islands in the archipelago (there are eighteen "main" islands in the Galápagos, and many more small islets and rocks that poke above the surface). Presumably these intrepid migrants came from the closest continental land, South America. With few or no predators in their new home, the finch population expanded rapidly. A few individuals traveled to other islands, either accidentally or in search of food or territory, and colonized them too. Although the distances involved for these "internal" migrations were not much more than fifty miles at the most, the journeys did require travel over water, and presumably they also involved only a small number of individuals, again ensuring an initially small and biased sample of the gene pool on each newly populated island. Gradually, through a combination of genetic drift, mutations, and natural selection, the finches evolved into distinct species that inhabited different islands and environmental niches. From an original small and presumably fairly uniform population, thirteen separate endemic species have evolved. The main differences among them are related to their diets and are primarily reflected in the astonishing array of bills the birds possess. Some have massive bills for cracking hard seeds, others small, sharp bills with which they peck ticks and other parasites out from the skin of iguanas and tortoises. Still others have bills adapted to manipulating cactus spines, which they use as tools to dig out insect larvae. One species uses its sharp bill to inflict small wounds on seabirds; it then drinks their blood. Even the most fertile imagination would be hard pressed to come up with the amazing array of feeding solutions that evolution has brought to these finches.

Birds, especially land birds, hold a special place in island biology. In part this may be because they have an easier path to initial colonization of, say, a newly formed volcanic island than do other plants and animals. Darwin's Galápagos finches are justly famous, but they are by no means the only birds on isolated islands or island groups that exhibit extensive evolutionary changes. On the Hawaiian Islands, for example, there are twenty-three endemic species of honeycreepers, all of which have evolved from a single, seed-eating ancestral species of finch. It is thought that even more endemic species of these birds—probably more than thirty—existed not so long ago, but colonization by humans has led to the extinction of some of them. Like the Galápagos finches, the primary distinction among the Hawaiian honeycreepers is feeding behavior, reflected in the shape of their beaks and tongues, which have variously evolved for feeding on seeds, fruit, insects, and nectar. But perhaps one of the most spectacular illustrations of island speciation among birds, and one that invariably catches the public imagination, is the birds of paradise. There are thirty-nine species, most with elaborate, colorful plumage, and all of them evolved from a common ancestor that resembled an ordinary crow.

Birds of paradise had been known in Europe for hundreds of years when *Challenger* set sail, but a comprehensive account had only appeared a few years earlier, in Alfred Russel Wallace's 1869 book *Malay Archipelago*. Wallace wrote that these birds were "the most extraordinary and the most beautiful of the feathered inhabitants of the earth." The *Challenger* naturalists were anxious to learn more about them and to collect as many specimens as possible. Their first opportunity came in early September 1874, not on a small island, but in Australia—at Cape York, to be precise, the village at the northern tip of Australia where Moseley found the aboriginal people to be in a "low state." There the *Challenger* scientists collected several specimens of the so-called "riflebird" (the Yidinji people of the region also know this bird by the wonderful name duwuduwu), a striking velvet-black bird of paradise with an iridescent blue-green bib and cap, one of several bird of paradise species native to northeastern Australia. Strangely, this first encounter with birds that the *Challenger* scientists had so much anticipated seeing merits only a few sentences in Moseley's notes and the journals of others on the expedition. But several weeks later, when the ship arrived at the Aru Islands, birds of paradise became the focus of attention.

Moseley noted that the Aru Islands, which lie about ninety miles south of New Guinea, are familiar to all naturalists from the work of Wallace, and are "so full of interest to us as the home of birds of paradise." This was not entirely accurate. The real home of the birds of paradise—the family of birds known as Paradisaeidae—is New

Guinea, a much larger island with a wide variety of habitats and consequently many more species. But the Aru Islands were famous for one species in particular, the greater bird of paradise. The male has spectacular plumage, and for generations the islanders had hunted the birds and traded their skins; many of these had made their way to Europe, and for that reason alone the *Challenger* naturalists would have been familiar with the birds. This was also the species Wallace had written about, and he had studied them in the Aru Islands. He may have been the first Westerner to observe their mating behavior; certainly he was the first to describe it. His observations gave him insight into the function of the birds' exotic plumage: to attract a mate. Wallace's work was recent and intriguing, and for that reason the *Challenger* naturalists made searching for the greater birds of paradise a high priority when they were in the Aru Islands.

The male greater bird of paradise is stunning, especially during courtship, when its plumage is most colorful. Its head is a palette of sharply contrasting colors: bright yellow, dark, almost iridescent emerald green, and deep, deep black. Its wings and back are a rusty brown, and on its flanks are masses of fluffy, whimsical feathers that rise up over its back like a sail during the courtship display. Like all birds of paradise, the female of the species is, by comparison, drab. But paradoxically, in this species and the others, it is the females who have driven the evolutionary processes that led to the spectacular male plumage. By choosing to mate only with males who mount the most dazzling displays, they have guided evolution toward ever more elaborate and sometimes astonishingly bizarre assemblages of feathers. The males themselves have had little evolutionary say in the matter. But in their courtship displays they make full use of what evolution has given them.

Wallace had been drawn to the Aru Islands because he was aware of the brisk trade in greater bird of paradise skins carried on by the islanders. Traditionally they had prepared the birds by removing feet and wings, a practice that they thought made the remaining plumage even more striking. Initially this confused some Europeans, who had never seen the birds in the wild and thought they might be some kind of magical, mythical creature: birds without wings or feet some-how floating eternally through the air in some far-off place. Possibly in Paradise. Their confusion was fairly quickly cleared up, but natu-ralists remained perplexed about the purpose of the unusual plum-age. When Wallace saw the mass courtship display by a group of male greater birds of paradise on the Aru Islands—multiple birds moving and bobbing among the branches, their fluffy, golden-yellow sails raised above their backs—he had a revelation. The extraordinary feathers, like the peacock's tail, were intended to attract females. Dar-win was tormented by the peacock's tail because he could not imag-ine what function it served; it seemed to defy his ideas about natural selection. The large tail did not keep the birds dry, or help them fly, or keep them warm. He thought that it might even be a hindrance to their survival, a clumsy appendage that restricted their mobility. But then he realized that peacocks' tails did have a purpose: they helped the birds mate. The best displays won. This was not natural selection in the way he had imagined it, but it was nevertheless se-lection: sexual selection. Among the birds of paradise, sexual selection has driven the evolution of even more extreme types of feathers than those of the peacock's tail, some barely resembling feathers at all. Researchers who study these birds are clear about how the develop-ment has come about, but they are still in doubt about why the fe-male birds prefer to mate with males who have the most elaborate, but hardly utilitarian, plumage.

When the *Challenger* naturalists arrived in the Aru Islands, none of them had seen greater birds of paradise in the wild, let alone ob-served the mating behavior that Wallace described. Moseley's first attempt to find one was frustrating. He was in the forest with native guides when suddenly a flock of birds flew right over his head; it was early morning and misty, and he did not realize what they were. When the guides told him they were greater birds of paradise he was annoyed that they had not alerted him to the birds' presence earlier. As always on such excursions, he had a gun with him—obtaining specimens for the *Challenger* collections was a high priority—but he had no time to shoot. He decided that the guides probably had a pro-

The greater bird of paradise, from *Histoire naturelle des oiseaux*, by François Le Vaillant (Paris, 1806). This black-and-white image gives a sense of the bird's beauty, but full appreciation requires seeing it in color. (Rare Book Division of the New York Public Library, courtesy of the New York Public Library Digital Collection.)

prietorial interest in preventing him from shooting the birds; if they killed them themselves they could sell the skins.

For a while the group pushed farther into the forest, following the birds' calls, but to no avail. However, a few days later, Moseley had better luck. This time his guides were more cooperative, possibly because he promised to pay them for every bird he shot. They quickly brought him to an area where the birds were calling and began to gesture and point up into the trees. Moseley admitted that at first he could not see anything. He did not have "native eyes," the acute awareness of their surroundings that people who live in nature possess and urbanized people have lost. His guides were experts with long experience, and they knew the habits and coloration of the birds intimately. But even when they pointed one out to him, Moseley would no sooner focus his attention on a particular spot than the bird was gone. They were constantly on the move through the foliage, flying from place to place rapidly and silently. When he finally did spot several of the birds they were high up in the trees, too far away for his shot to be effective. Finally a female flew up from the ground nearby, and (as cruel as it seems to us today) he immediately shot her. At last he had a specimen. He was even more eager to bag a male in full plumage, but it was not to be. He saw—or thought he saw—just one during his time on the Aru Islands. He caught only a fleeting glimpse of the bird, and afterward he began to wonder whether his imagination had been playing tricks on him. Perhaps it had not been a male in "full dress" after all. The native guides told him the best time to capture the birds was during the mating season, when they congregated in places known to the islanders and were too busy attracting females to pay much attention to human intruders. But *Challenger* was gone long before the mating season arrived.

There is a passage in Moseley's notes from the Aru Islands that sheds light on his personality and his sense of humor—although it also highlights once again the yawning gap between his time and our own, as well as the sense of their own superiority that the *Challenger* scientists and Europeans in general carried with them, consciously or not. On his shore excursions Moseley always took with him one or two local guides. One whom he hired in the Aru Islands was particularly active; the man sped through the dense, hot jungle so rapidly that Moseley had difficulty keeping up, to say nothing of making any useful observations. But because he did not want the guide to think he was lazy or unfit, he stopped to examine a large stone, turning it over and inspecting it carefully. He indicated to the guide that it was of great interest, something to be kept for his collection. For the rest of the day the poor man was burdened with the heavy "sample," slowing him considerably and permitting Moseley to travel at a more comfortable pace. (He used this ruse on other occasions, too, and in his journal he recommended it to naturalists who found themselves in similar circumstances.) At the end of the day, he writes, "I conveyed the stone on board the ship with due solemnity and threw it overboard."

Although his quest to obtain a male greater bird of paradise was unsuccessful, Moseley was more fortunate with another species, the king bird of paradise, or "King-Bird" as he called it. This smaller species was much more common in the Aru Islands than the greater bird of paradise, but the males were equally spectacular. Their heads, backs, and sides are a brilliant cardinal red, their breasts snowy white, topped with a coal-black band across their throat that gives them the appearance of wearing a scarf. Their feet are bright blue, and on their shoulders green-tipped feathers open up like a fan during the courting display. Topping it off are specialized feathers that do not resemble feathers at all: two long "tail wires" tipped with small green disks. Tail wires are a feature of several other bird of paradise species as well, and all those that possess them wave them about in elaborate displays during courting. Moseley observed that the king birds, like the greater birds of paradise, were in frequent motion through the forest but appeared to be quite tame. "I shot five of them in one day," he wrote, and added, "Luckily, the skin of the Paradise Birds is quite tough, and I found the King-Birds easy to skin."

In the Challenger Report, the section dealing with birds lists a total of fourteen bird of paradise specimens collected from the Aru Islands: five greater birds of paradise, seven king birds of paradise, and two specimens of another much less gaudy species that is a glossy green-blue-black color and looks more like a small crow or a starling than an exotic bird of paradise. These specimens were in addition to six riflebirds collected at Cape York, four of them males in their full plumage. Moseley was not alone in pursuing these birds; the other naturalists, as well as the ship's officers, joined in the hunt. All were preoccupied with finding greater birds of paradise, presumably because of Wallace's account. To their disappointment none of those they shot were males in full plumage.

Unfortunately, the *Challenger* scientists were not able to continue their search for birds of paradise in New Guinea, which, as noted earlier, is home to the most diverse collection of these beautiful creatures. The ship's only stop there, at Humboldt Bay, was where Moseley came face to face with a native trader who drew back his bow and pointed the arrow directly at his chest. Given this reception the stay was brief, and the naturalists had no opportunity to trek inland in search of specimens. But why is it that the birds of paradise are found in such variety in New Guinea? The story begins more than a hundred million years ago, when New Guinea as we know it was part of Gondwana, the ancient supercontinent that was made up of all the southern continents—South America, Africa, India, Antarctica, and, crucially, Australia. However, on geological timescales nothing is permanent on this earth, and Gondwana eventually broke up, the present-day continents drifting apart until they reached their current positions. They are still drifting and will continue to do so; a world map from a few hundred million years in the future would be unrecognizable to us today.

As the supercontinent of Gondwana broke apart, Australia was the last large piece to split off from Antarctica. The separation happened gradually, from west to east along what is now Australia's southern coast, and between thirty-five and forty million years ago the last land connections to Gondwana were severed. Australia became an island—a gigantic one, but an island nevertheless. Its subsequent isolation, with no land connections to other continents, has had enormous consequences for the evolution of its flora and fauna and is the reason its native plants and animals are so different from those of the rest of the world. Geologically, New Guinea and Australia are both part of that same large piece of continent that split off from

Gondwana. The reason we recognize them as separate entities today is that a shallow sea covers part of the continent between them; in the past, due to fluctuations in sea level, dry land connected the two. This past connection accounts for the close similarity of their flora and fauna: there are kangaroos in New Guinea and birds of paradise in Australia. Current research indicates that birds of paradise actually originated in mainland Australia, probably around twenty-four million years ago. Today, as mentioned, only a few species are native there; they are restricted to the northeastern part of the country. But in New Guinea the Paradisaeidae proliferated and diversified. The complex geology of the island has provided an abundance of ecological niches for the birds, from high mountains to rainforests and flat coastal lowlands. New Guinea's isolation has limited the arrival of invasive species, and until humans appeared the island lacked large predators. An abundant supply of fruit and insects meant that nourishment posed no problem. The birds spent relatively little time and energy foraging; this may be one of the reasons they could devote themselves to perfecting their courting rituals.

It is only in recent years that many of these displays have become well known. In 2004, Tim Laman, a biologist and wildlife photographer, teamed up with another biologist, Edwin Scholes, and set out to photograph birds of paradise for *National Geographic*. Their project evolved into a monumental quest that took eight years to complete, but by its end they had photographed every one of the thirty-nine existing species. No one had ever done that before. In the course of their work they also captured the first video footage of many of the birds' elaborate mating behaviors. Some of these videos, as well as others taken a few years later by BBC cameramen for David Attenborough's television series *Planet Earth*, are freely available for viewing on the internet. They are worth watching, and I provide links to some at the end of this book. Birds dancing, birds transforming themselves into shapes that do not resemble birds at all, birds tidying up their "stage" before performing, birds waving their elongated tail wires and head feathers in routines that look like ribbon gymnastics. Not to mention birds strutting, hopping, puffing up, and flashing displays of iridescent color. It is no wonder Wallace was entranced.

If birds of paradise were elusive for the *Challenger* naturalists, other brightly colored creatures on the Aru Islands attracted their attention. On one of his forays into the bush Moseley observed an eye-catching group of male butterflies with "brilliant green and velvety black wings" fluttering around a single female. They were, he assumed, competing for her attention, and this gave him an idea. It would be possible, he wrote, to investigate the impact of butterfly coloration on mating behavior by doing some simple experiments. He could rub the bright colors off their wings, or paint them over with a dull color. Transformed, would they still be able to lure a female? Moseley was always searching for answers. He thought that he could do similar experiments with "gaudy birds." Since he wrote these notes in the Aru Islands, he was probably thinking of birds of paradise. Brilliant feathers could be painted over with muted colors, or a male not yet in full plumage could be painted with bright colors. He might try out different colors to see which were most effective. As far as he was aware, no one else had done systematic investigations in this vein before, and in his published journal he writes enthusiastically about the possibilities. But such experiments were impractical during the expedition, especially with live birds, and at any rate Moseley had much else to keep him busy. There is no evidence that he followed through on his idea when he returned to England, either.

He did, however, collect specimens—as many specimens as possible. All the naturalists did. It was one of their primary goals, and they collected with a passion, even—or perhaps especially—when they knew a species was relatively rare. This was the way natural history was done in their day, and collecting is a recurring theme in all the accounts of the *Challenger* expedition: the scientists' job was to gather as many examples as possible of every plant, every animal, every rock and mineral from the seafloor, and then examine, describe, and catalogue them. A new species or a new type of material from the ocean depths was a prize, something that could be added to the existing fund of knowledge. For the most part none of the naturalists or others on the expedition questioned this approach, at least not in their journals or other records of the expedition. Recall Moseley's comment about the beautiful king birds of paradise: "I shot five of them in one day." He betrays no hint of contrition. But occasionally in his writing, and also in that of George Campbell, a suggestion of remorse appears, or at least a sense that if animals had to be killed, it should be done judiciously. In the Kerguelen Islands, Moseley records that he killed two fur seals, while others from *Challenger* killed an additional two. These were for science, and thus acceptable. However, he complained that the commercial sealers killed every seal they could find, and he feared that such wholesale slaughter would cause the population to diminish rapidly. "It is a pity that some discretion is not exercised in killing the animals," he wrote. He noted that in some other places, where only carefully selected animals were harvested, the populations actually increased.

Moseley had another experience on Kerguelen that gave him pause. One day he came upon a group of teal—and he shot them. This time, however, he had not acted for science. The ducks made good eating, and there were many men to feed. He described how the tame and unsuspecting birds waddled right up to him, and it is clear he was not proud of what happened next. "I hesitate to describe the termination of the scene," he wrote. But, in self-defense, he continues: "Only those who have been long at sea know what an intense craving for fresh meat is developed by a constant diet of preserved and salt food." Campbell was even more contrite after he took part in a kangaroo shoot in Australia. He wrote that for some reason he had had an "insane desire" to shoot a kangaroo, and friends who owned an inland farm took him on a hunt to do just that. But instead of elation, the experience gave him only a feeling of disgust. The dead animals were left lying forlornly in the fields, with the exception of their tails, which were used to make soup. Describing the excursion Campbell confesses, "I must tell you that after I had shot that object of my ambition—a kangaroo—I felt no sort of exultation; on the contrary, I thought the proceeding a mean and unsatisfactory act."

Museums in Victorian times were typically crammed with as many specimens as possible, as is evident in this photograph from the Anatomy Museum of the University of Edinburgh, taken in 1898. (Courtesy of the Centre
for Research Collections, University of Edinburgh.)

Hunting for food or for "sport" aside, the quest for specimens to
take home and examine was unrelenting, even if the *Challenger* scientists occasionally had reservations. Victorian-era museums reflect
this mania for collecting; if there is a word that sums them up it is
abundance. Photographs from the period show display halls crammed
with herds of skeletons, display cases packed with flocks of birds
and butterflies. It was important to have and exhibit as many examples and variations of a thing as possible. The *Challenger* naturalists
were imbued with this ethos. Although precise details had not been
worked out beforehand, they understood that the vast collection of
specimens they expected to accumulate would in the end be housed
in a museum.

Some scholars mark the decade of the *Challenger* expedition, the 1870s, as a turning point in how knowledge about nature was gathered and transmitted, especially in Britain. Museums, as both repositories of objects and, increasingly, places of research, reflected this change. Especially in the wake of Darwin's evolutionary theory, the focus of scientific investigation was shifting from form to function, from the visible characteristics of a plant or animal to the less obvious reasons for its appearance or morphology or geographical distribution. There had always been curious scientists pondering why things were as they were. But in biology the theory of evolution provided a new framework for asking questions about purpose. The peacock's tail that perplexed Darwin is a good example. It did not appear to fit the framework because it did not seem to offer anything that would enhance its chances of survival. But if so, why was it there?

Two very different museums in Victorian Britain, both with connections to the *Challenger* expedition, give a taste of the changing nature of natural history studies. John Murray's grandfather John Macfarlane built his natural history museum in the Scottish town of Bridge of Allan in part as a personal project, a passionately pursued hobby to occupy his retirement years, and in part because he believed it would be an instrument of public education and self-improvement. Its main purpose was display. He was a collector, and clearly the process of pursuing and obtaining specimens was a thrill for him in its own right. He wanted the collection to be as large and comprehensive as possible, to rival those of other, more famous museums. He was not a biologist and he did not personally make collections in the field; rather he accumulated specimens from every possible source: by purchase, through his business contacts, by way of his grandson's collections of local flora, fauna, and rock samples, and through a friend who was curator of a natural history museum in Manchester. The larger and more exotic the animal, the keener he was to obtain it for his museum. He ordered a Bengal Tiger, lions, an ostrich, a boa constrictor, an elephant. He made some attempt to group specimens according to type—mammals in one place, birds or minerals or shells in others—but there is no hint that research would play a part in

Macfarlane's museum. The young John Murray, appointed by his grandfather to curate the museum, imposed some order in the displays. But for a visitor, the museum experience was meant to be visual, part learning and part entertainment, a way to become acquainted with the wonders of the natural world.

As Murray was gaining experience in his grandfather's museum, changes were afoot in the much larger London museums. In the capital, natural history collections were housed in the British Museum, which was also home to everything from Egyptian mummies to ancient coins and stone carvings from Indian temples. Many felt that natural history got short shrift. The museum did not have enough space; existing specimens were crowded together in the displays, and new ones were constantly arriving, donated by collectors or sent back by scientists exploring the distant lands of the expanding British Empire. Eventually the government was persuaded to fund a new museum dedicated to natural history. Construction began in 1873, the first full year of *Challenger*'s voyage, and the museum's curators were already lobbying to gain control of the collections that would eventually be returned to Britain from the expedition. Richard Owen, a biologist who was the new museum's most persistent promoter and its first director, wanted the building to be large enough to display every species of each class of organism. And even within a species he felt that as many as possible of the subtle varieties that existed should be laid out for all to see, to allow the public to observe the true diversity of living things. This kind of thinking also drove the efforts of the *Challenger* naturalists to collect as many specimens as they could. In Owen's writings he notes that such an approach was important in a museum "destined to gratify the curiosity of the people, and afford them subjects of rational contemplation." The new natural history museum was to be for the people. But it was also to be a place where scientists could examine and contemplate the variety displayed in the collection. Like London's new Natural History Museum, other museums around the world were becoming centers of research as well as of spectacle.

It was always understood that the *Challenger* collections were primarily destined for research; at the conclusion of the expedition all of the non-marine specimens were transferred to the Natural History Museum for study by specialists. Meanwhile, the marine samples were kept at the Challenger Office in Edinburgh, either to be studied by those associated with the voyage or to be distributed to other experts for further investigation. But when the Challenger Office eventually closed these marine materials were also sent to the Natural History Museum; today it holds over ten thousand specimens from the voyage—a valuable historical archive for researchers.

Owen's Natural History Museum is now a center for scientific investigations, with hundreds of scientists on its permanent staff. It is also one of London's top attractions, welcoming millions of visitors each year to its remarkable displays, displays that still have the ability, in Owen's words, to gratify the curiosity of the people. Some of the exhibits include samples brought back by the *Challenger* naturalists. They are, one hopes, suitable subjects for "rational contemplation."

eleven LIGHT AND BLACKNESS
IN THE DEEP SEA

The *Challenger* scientists called it phosphorescence. Today it is better known as bioluminescence, and if you've ever watched fireflies flashing through the air on a summer night or stumbled upon a glowworm in the dark, you've experienced the phenomenon for yourself: living organisms, generating their own light. Bioluminescence is more common in the sea than on land; it has been estimated that at least nine of every ten species that live in the deep ocean have the ability to luminesce. The U.S. Postal Service issued a series of stamps celebrating bioluminescence in 2018; eight of the ten stamps depict marine organisms. Growing up, I was familiar with—and like most people fascinated by—fireflies, but my first experience with marine bioluminescence came on a September night on the Oregon coast. We were camping, and after setting up our tent and eating we walked along the sandy beach. Close to the water's edge, where the sand was still damp from the waves, every step generated a radiating halo of tiny pinpricks of light spreading out from our feet. It was magical.

The phenomenon of bioluminescence has been written about since ancient times, but even in the 1870s little was known about how it occurred. A short article that appeared in the journal *Nature* not long before *Challenger* embarked on her journey is instructive. Today *Nature* is one of the world's premier scientific journals, full of detailed, often complex articles discussing cutting-edge research. In the 1870s it also provided a forum for discussion of natural phenomena. A letter from one John James Hall that appeared in the October 3, 1872, issue was titled "Phosphorescence in Fish." It immediately followed a short note about the significance—or lack thereof—of an extra tooth in cats. At any rate, Mr. Hall (most likely the same John James Hall who was a prominent British clockmaker, restorer, and author, although it is impossible to know for sure as only his name appears in the journal) had been on a ship off the southwest coast of England a few months earlier and had observed "one of the most beautiful marine phenomena that could well be imagined." Large patches of the sea were luminous. It was a rough night, the ship rolling and pitching, and to get a closer view he "secured [himself] by a tight hold on the stanchions immediately over the bow." The eerie light, he concluded, came from "the phosphorescence of immense shoals of fish." But the phosphorescence was not restricted to the fish; when a wave broke on the bow the drops of spray that landed on him and the deck shone too, like "so many glow-worms." Hall thought that the luminescence of the water must somehow originate with the fish, probably through a secretion they gave off. He referred to work of Paolo Panceri, a professor at the University of Naples, who had written extensively on phosphorescence. (Bioluminescent organisms are abundant in the sea around Naples.) Panceri had recently concluded that phosphorescence in fish arose from fatty material "cast off by the animal—it is a property of dead separated matter, not of the living tissue." This idea probably stemmed from the observation that dead and decaying fish—more common in the days before refrigeration than today—are sometimes luminescent. But we now know that this particular natural light is produced not by the fish but by bacteria living in their dead flesh.

The crew of *Challenger* observed light displays at sea that were even more spectacular than the one described by John James Hall. When the ship left the Cape Verdes Islands in early August 1873, every night seemed to bring a more dazzling show of bioluminescence. George Campbell, who had sailed the world with the navy and had presumably many times before seen "burning seas" or "milky seas," as the phenomenon was sometimes called, summoned up his most expressive prose to describe one of these displays:

> On the night of the 14th the sea was most gloriously phosphorescent, to a degree unequalled in our experience. A fresh breeze was blowing, and every wave and wavelet as far as one could see from the ship on all sides to the distant horizon flashed brightly as they broke, while above the horizon hung a faint but visible white light. Astern of the ship, deep down where the keel cut the water, glowed a broad band of

blue, emerald-green light, from which came streaming up, or floated on the surface, myriads of yellow sparks, which glittered and sparkled against the brilliant cloud-light below, until both mingled and died out astern far away in our wake. Ahead of the ship, where the old bluff bows of the Challenger went ploughing and churning through the sea, there was light enough to read the smallest print with ease. It was as if the milky way, as seen through a telescope, scattered in millions like glittering dust, had dropped down on the ocean, and we were sailing through it. That is, if you will, a far-fetched comparison; but a more or less true one all the same.

Of course the *Challenger* naturalists wanted to investigate, so they trawled frequently and took samples of the scintillating seawater. They found a huge variety of bioluminescent organisms. The main source of the light during the first few nights after leaving the Cape Verdes seemed to be a large, cone-shaped gelatinous organism that the scientists knew as *Pyrosoma*, which is actually not a single creature at all but a colony made up of a myriad of tiny individual animals. Wyville Thomson wrote that *Pyrosoma* gave off a white light "like molten iron." On the night described by Campbell, other small floating organisms that the naturalists thought were a type of diatom were responsible for the display. These were tiny and single-celled, but there were millions of them, enough to light up the ship. Later, when they were examined in more detail, it was found that they were not diatoms but a new species of dinoflagellate, the same planktonic organism that causes so-called red tides.

The *Challenger* naturalists encountered surface-dwelling bioluminescent organisms throughout the voyage. It was a phenomenon they expected to see, especially in the tropics, although only rarely did the light shows rival those they witnessed early in the expedition. But something they were not prepared for was the prevalence of luminescent species in their dredge hauls from the *deep* ocean, although Wyville Thomson had already had a hint of this possibility from his pre-*Challenger* deep dredging in the seas around the British Isles. In *Depths of the Sea* he describes luminescent starfish and other creatures dredged from depths of thousands of feet. He also notes that in some of the same dredges he found other organisms with well-developed eyes. This was far beyond the depth to which sunlight could penetrate, and he wondered why these creatures had eyes if there was no light. Was it possible that the whole of the deep sea was lit by the low, eerie light of luminescent animals? It was a question, he wrote, of "extreme interest and difficulty," one that would require much additional investigation. Thomson's idea caught the imagination of many scientists, and for a while it was known as the "theory of abyssal light."

Nearly all the deep sea corals dredged by *Challenger* were bioluminescent, a finding that intrigued Moseley. He was curious about the purpose of bioluminescence and concluded that for the corals, at least, it probably was not meant simply to light up their surroundings because related species living in shallow, sunlit water were also luminescent. Clearly these organisms had no need to illuminate their environment. Perhaps, he speculated, light emission was a by-product of some other process, analogous to the heat produced by human bodies. He had a vivid imagination. "It is quite conceivable," he theorized, "that animals might exist to which obscure heat-rays might be visible, and to which men and Mammals generally, would appear constantly luminous." Who would have thought that a century later humans would be donning night-vision devices so they could see others in the dark by the "heat rays" they emitted.

Like Thomson, Moseley was fascinated by the presence of eyes in some of the animals dredged from great depths. Again he tried to imagine their purpose. Perhaps, he thought, the animals with eyes congregated around patches of luminescent corals on the seafloor. His words paint an image of fish and other creatures hovering around a dim blotch of light on the ocean bottom like campers around a fire on a dark night. Moseley also wondered whether the colors of deep sea organisms might be tied to bioluminescence—the light has a limited spectral range, so only certain colors would be visible under its illumination. What use would a color have if it could not be seen? (He apparently never considered camouflage.) He tried to investigate the spectrum of the light given off by several different species of luminescent deep sea corals, and according to his spectroscope—crude by today's standards—it contained red, yellow, and green but was missing the blue part of the spectrum. Modern work shows that most marine bioluminescence is limited to blue-green wavelengths, so it is unclear whether he made an error in his measurements or was working with unusual species. Moseley concluded, though, that if the luminescence spectrum lacked blue, there should be no blue organisms in the deep sea, and this agreed with his observations from *Challenger*'s deep dredge hauls. But in his quest to understand the colors of deep sea organisms he was unable to come to a definitive conclusion, for he knew that blue is not a common biological color. That, not the details of deep sea luminescence, could be the reason for the lack of blue animals in the dredge hauls.

One of the most consequential discoveries about bioluminescence was made a decade after the conclusion of the *Challenger* expedition by the French scientist Raphaël Dubois, who showed that the light is produced by a chemical reaction choreographed by the organisms and is not a by-product of some other process, as Moseley had speculated. In his research Dubois worked with a pair of luminescent animals, a beetle and a clam. The clam had long been famous because of its ability to light up the mouths of people eating it. He isolated two chemical compounds from the luminescing organs of these animals and found that when they were combined, they produced light. He named the compounds luciferin (light-bringer) and luciferase; luciferin is the substance that produces the light, while luciferase is an enzyme that enables the reaction to occur.

Dubois's work was important because it showed that bioluminescence is a process that takes place in living cells. It is different from other kinds of phosphorescence that require external stimulation—for example, by exposure to sunlight. Glow-in-the-dark objects, like the T-shirt I have with a glow-in-the-dark star map on the front, are not bioluminescent; they only shine after they have been stimulated by exposure to light. But decades after Dubois's work, bioluminescence continued to be enigmatic. In 1931 another note in the journal *Nature* about phosphorescence at sea concluded: "Nothing is in reality known about this mysterious phenomenon, save that it is associated with floating animals, and possibly at times with floating plants."

An indication of the importance of bioluminescence is that the phenomenon appears to have evolved independently in different groups of organisms (both marine and terrestrial) forty or fifty times over the past few hundred million years. That light production has developed so many times in so many different species illustrates that it must provide evolutionary advantage or usefulness. Long before Moseley's day it was known that among fireflies—perhaps the most easily observable bioluminescent creature for early investigators—the light was used to attract a mate. Like bird of paradise plumage, it is a product of sexual selection. When Moseley suggested that deep sea luminescence might be an accidental development, he hedged his bets by adding, "although of course, in some cases, it has been turned to account for sexual purposes, and may have other uses occasionally." Research in recent decades has provided much more detail. To return for a moment to the firefly, it has been discovered that its light signals, with long flashes, short flashes, and sometimes a distinct sequence of flashes, are a kind of insect code, specific to the species and the sex of the sender. Even the color of the light, its precise frequency, may convey information to the observer. One type of North American firefly mimics the flashes of other species to attract an unsuspecting "mate." But it's not reproduction she's interested in, it's her next meal. When the hopeful mate arrives, she eats him.

Over much of the volume of the ocean bioluminescence is the only source of light, and for many marine creatures it is the most important form of communication. Recent research has focused on the genetic makeup of bioluminescent organisms and on the specific chemical compounds that produce the luminescence, and through this work researchers have been able to trace how different groups of animals have independently developed bioluminescence capabilities through geological history. Scientists have also characterized in more detail the chemical reactions involved in biological light production. The research led to a Nobel Prize in 2006 for the marine biochemist Osamu Shimomura, who identified and isolated a now famous (among biochemists, at any rate) fluorescent protein from a bioluminescent jellyfish. Shimomura shared the prize with two others, who developed ways to use the protein as a tag to follow complex biological processes.

But sorting out the functions of bioluminescence—why and how it is used—is even more difficult than isolating a fluorescent protein. Functions can rarely be measured directly; they have to be inferred. Field observations are often crucial, and these can be difficult to make on marine organisms, particularly those that live in the deep sea. Nevertheless, marine biologists have recorded an amazing diversity of behaviors that involve luminescence. In the simplest sense they can be grouped into a small number of use categories: communication, offense, and defense. But organisms use their self-produced light in multiple ways within those basic groupings. And some creatures employ their bioluminescence capabilities across all these categories.

Take, for example, the so-called "flashlight" fish. These are small fish that live in the tropics, and they often feed at night in the shallow water of coral reefs. Their behavior has been studied in some detail both in the field (by scuba-diving biologists) and in the laboratory. The fish are not themselves bioluminescent, but underneath each eye they have a crescent-shaped organ full of symbiotic, brightly luminescent bacteria that produce light continuously. The fish can open or close a dark, eyelid-like structure over this luminous organ, effectively turning its "flashlights" on and off at will. Flashlight fish have been observed to use the light from their bacterial partners for all three of the purposes mentioned above: offence, defense, and communication. During nighttime hunting they really do use the luminescence like a flashlight, turning and rolling their bodies so that the light beams sweep through the water in search of prey, which is usually a variety of small crustacean that is attracted to light. Sometimes the fish swarm in schools with their flashlights turned on, creating large patches of light in the dark ocean and drawing in even more of their prey. They are also skillful at using luminescence for defense. One of their observed maneuvers has been called "blink and run": to avoid predators the moving fish shuts the lids on its luminescent organs, abruptly changes course, and then opens the lids to turn on the lights again only when it is in a different location and swimming in a different direction. This is an effective feint that nearly always confuses a predator. Finally, observations in the laboratory have revealed that fish kept in separate tanks blink back and forth at one another in Morse code mode, using their bioluminescence to communicate. No one has yet deciphered their code; what they are saying is unknown.

Challenger's deep dredge hauls regularly brought up fish, many of them luminescent. There was always a nagging question of whether they were truly from the bottom or had been caught up in the dredge somewhere between the bottom and the surface—a question that did not arise with organisms such as corals that were known to live on the seafloor. Most of the dredged-up fish, though, had physical features suggesting that they indeed lived at great depths. The majority were also new species. When the expedition began only about thirty deep sea fish species were known; by its end the tally had increased by an order of magnitude to more than three hundred. Many were badly damaged when they reached the surface, the result of sudden decompression from the high pressures under which they lived. But in spite of their poor condition, some were still luminescent. Rudolf von Willemoes-Suhm wrote that one fish brought up at night "shone like a star." Many that were not obviously luminescent had "organs" on various parts of their bodies that the naturalists took to be sites of bioluminescence. In the Challenger Report the section on deep sea fish has two appendixes devoted to descriptions of these unique structures. One is by Henry Moseley. The other is by a biologist from University College London, Robert von Lendenfeld. The reports emphasize the degree to which—if you'll excuse the expression—these scientists were still working in the dark as far as bioluminescence is concerned.

Although Lendenfeld had little understanding of the way the deep sea fish produced their light, his report is a remarkably thorough descriptive account of their varied and sometimes extremely complex bioluminescent organs. Today the structures are known as photo-phores; Lendenfeld calls them "glandular organs," or "phosphores-cent organs," and he describes their diversity: from small round objects filled with cells of various types to more elaborate structures with reflectors, layers of pigment, and even lenses. He notes that they occur in different places on the body in different species, and he ties their location to what he thinks is their probable function: light or-gans on the head and near the eyes were probably used to search for or lure prey; organs on the back or belly might scare away predators. He describes the nerves and blood vessels that connect the organs to the circulatory and nervous systems of the animal, and concludes that the fish must be able to control their displays of luminescence, although he does not know how. Dubois's work on luciferin and lu-ciferase was still to come, and Lendenfeld thought that the light was probably generated by "luminous slime." The luminescent organs, he notes, had a "special phosphorescent apparatus" that produced "light at the volition of the fish by using up or burning the secretion by the gland." But he also injects a note of caution. He cannot be cer-tain that all the structures he studied were genuine phosphorescent organs because some of the fish were not visibly luminescent at the time of their capture. Nonetheless, his investigations indicated that most of them were designed to produce light. Even though some had lenses and other features usually associated with eyes, he was reason-ably certain the "glandular organs" were not used for sight.

Moseley's contribution in the Challenger Report comes to the same conclusion. It is less comprehensive than Lendenfeld's; it deals with only a single species. But it focuses on the question of eyes ver-sus luminescent organs. John Murray had retrieved a peculiar-looking fish from a dredge and he had been intrigued by the pair of strange, eyelike structures on its head. Once more the question arose: Why would a fish that lived in the dark have eyes? Murray concluded that that the odd features were not sight organs, but he asked Mose-ley to investigate further. Moseley made thin histological sections in order to study the tissues of the organs under the microscope, and

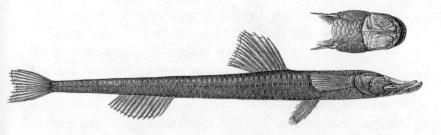

The strange deep sea fish *Ipnops murrayi*, dredged from a depth of 11,400 feet in the Atlantic Ocean, about 400 miles to the west of Tristan da Cunha. This specimen was about 4½ inches long. Note the odd, flattened eye structures that both Murray and Moseley initially thought were "phosphorescent organs." (Courtesy of the Centre for Research Collections, University of Edinburgh.)

his report is long, detailed, and accompanied by intricate drawings of the structures at high magnification. He concludes that Murray was correct, writing, "The peculiar organs have in reality no connection with organs of vision"; they are "most probably phosphorescent." But his initial assessment turned out to be incorrect. He later reexamined his histological sections and decided that the "peculiar organs" could, after all, be sight organs, designed to detect very weak light if not to discern objects clearly. Subsequent work on this fish—later named *Ipnops murrayi* after Murray—has shown that the structures are indeed eyes, and like those of most deep sea fish they are highly modified. How they are used in the darkness of the deep sea is still unknown.

Most deep sea fish are bioluminescent. The diversity of luminescent organs and the strategies of both fish and other marine organisms for using them are spectacular, in their own way rivaling the array of highly specialized plumage among the birds of paradise. The multipurpose headlights of the flashlight fish are just one example. Some marine organisms have photophores on their undersides that adjust to match light coming from above; when they are near the surface they become all but invisible to predators below. The vampire squid has photophores covering much of its body, and when in danger it emits disorienting flashes of light of varying duration to confuse predators. From the point of view of a predator it must be a bit like being blasted with strobe lights. If the squid is in extreme danger it goes even farther, generating a smokescreen by discharging a glowing, luminescent cloud of mucous from the tips of its arms. Small, shrimplike crustaceans that live on the seafloor similarly confuse predators by ejecting both luciferin and luciferase from tiny nozzles, creating a cloud of light in the water. Some squid shed luminescent body parts as decoys, like a lizard shedding its tail. Others have photophores arrayed along their dangling arms that act as lures to attract prey. There are jellyfish with luminescent tentacles; victims attracted to the light are immobilized by the organism's stinging cells. The anglerfish has evolved its own fishing rod: a modified back fin that projects out over its mouth and has a bioluminescent lure at the end. Much like the luminescent organ of the flashlight fish, the dangling bulb is not a true photophore but an organ full of symbiotic luminescent bacteria. It serves a dual purpose: female anglerfish use it to entice male partners, and both sexes use the light to attract prey. Anglerfish were known to the *Challenger* scientists, but until they dredged one up from a depth of nearly three miles in the Atlantic they had not realized the fish could live in such deep water. It was a new species, and when Wyville Thomson wrote about the specimen he observed, "It is the habit of many of the family to lie hidden in the mud, with the long dorsal filament and its terminal expansion exposed. It has been imagined that the expansion is used as a bait to allure its prey, but it seems more likely that it is a sense-organ, intended to give notice of their approach." It was one of the few deep sea fish that neither he nor his colleagues realized was bioluminescent.

Occasionally the *Challenger* scientists had moments of levity at the expense of bioluminescent marine organisms. Once a trawl net brought up a particularly large *Pyrosoma* (the jellylike colonial organism described earlier), and the naturalists put it in a tank on deck.

That night, Moseley says, he wrote his name on it with his finger. A few seconds later his signature appeared "in letters of fire."

Bioluminescence is a source of light in the deep ocean; manganese, or more accurately manganese oxide, the form in which the element occurs on the seafloor, is black and absorbs light. The *Challenger* scientists found manganese everywhere in the deep sea. I searched for the word *manganese* in the volume of the Challenger Report that deals with deep sea sediments and found it was used more than a thousand times. The first hint of the importance of manganese came early in the expedition, in February 1873, shortly after *Challenger* left Tenerife. Soundings indicated the ship was over a shallow bank that the scientists assumed was a volcanic structure connected to the nearby Canary Islands. Dredging brought up pieces of black rock, which seemed to confirm their assumption. A few rock pieces had branching coral attached; the coral was dead, and, like the rocks, it was jet black and shiny. At first the naturalists suspected that the coloration was caused by a coating of carbon. A few days later a dredge in much deeper water brought up red clay; scattered through it were small, peculiar-looking black nodules that were obviously neither plants nor animals. John Buchanan thought their black color, like that of the coated corals, might be due to carbon. As the expedition's chemist, he "considered that [his] department had to render an account of them [the nodules]," and he proceeded to do a chemical analysis. He was in for a surprise: his tests for carbon were negative. Instead, he found that the nodules were rich in manganese. And he got the same result when he belatedly analyzed the black coatings on the coral and the rocks the coral had grown on. The discovery that they were coated with manganese, and the visual impact of the dredge full of shiny black rocks and coral fragments, remained with him: "It was a sight which I see now before me as clearly as I did when on the deck of the *Challenger*," he wrote years later.

As the expedition progressed the *Challenger* naturalists soon became familiar with the odd, black, roughly spherical objects they found in the red clay, and before long they were referring to them as manganese nodules. The convention stuck; the nodules have been called manganese nodules ever since, even though in reality they are complex mixtures of manganese and iron oxides and contain about as much iron as manganese. (The word *manganese* is commonly used as a short form for "manganese and iron oxides and hydroxides," the forms in which manganese occurs on the seafloor. The *Challenger* scientists employed that usage, and I will follow the practice here.) Buchanan's chemical analyses showed that other metals were also present in the manganese nodules, although in minor amounts: he detected nickel, copper, and cobalt.

Especially in the deepest parts of the ocean, the *Challenger* scientists found nodules in most of their dredge hauls, and they also discovered that manganese was ubiquitous on the seafloor in other forms: as a crust on rock fragments, and as a thin coating on everything from shark's teeth to whalebones, pieces of pumice, the shells of foraminifera, and dead coral fragments. Like the nodules, the coatings are approximately equal mixtures of manganese and iron oxides. Manganese in all its varieties was so common that the Challenger Report states simply, "To mention all the regions where manganese was observed would take up too much space," and notes that only special occurrences would be discussed. The nodules, the most striking manifestation of manganese in the deep sea, merited their own section. Some of the *Challenger* dredge hauls in the Pacific came up bulging with them. When one such haul was dumped out on deck Murray likened them to a bunch of cricket balls. In recent decades sea-bottom photographs have shown large expanses of the ocean floor almost completely covered with them, looking for all the world like pictures of a field littered with newly harvested potatoes. The photos also show that the nodule concentration varies tremendously from place to place, sometimes over short distances. This makes it difficult to estimate total amounts, but the most careful assessments suggest that at least 500 *billion* tons of manganese nodules lie on the floor of the world's oceans.

The ubiquitous nodules and coatings raised a number of vexing questions for the *Challenger* scientists. Where did the manganese come from? What process could create a round nodule in an area of soft, flat, seafloor sediment? *Why* do they form? Did the nodules differ in some way from the manganese coatings and encrustations? The naturalists were not entirely without clues. For example, they knew that in the presence of even a small amount of oxygen, manganese would react to form oxides like those they found in the nodules. Seawater contains dissolved oxygen, so it made sense that any manganese that made its way into the ocean in a dissolved state would precipitate out as an oxide. Whenever the scientists cut open a nodule they invariably found a nucleus at the center; a shark's tooth, a small fragment of whalebone, or a piece of pumice were common. This indicated that a hard surface was necessary to initiate the precipitation of the iron and manganese oxides. Often the nucleus appeared to be highly altered from its original state, a clue that it was probably very old. And most of the sliced-open nodules exhibited closely spaced concentric bands around the nucleus, resembling growth rings. Taken together, these observations suggested that the nodules grew extremely slowly.

But if nodule growth was slow, there was a problem: Why were they not covered with the sediment that was constantly raining down on them? The *Challenger* scientists could think of only one possibility: the nodules were moved around by either bottom currents or burrowing animals frequently enough to keep them on top of the mud.

Equally perplexing was the source of the manganese. Its concentration in seawater is very low, and the analytical techniques available to nineteenth-century chemists were not sensitive enough to detect it. This led the *Challenger* naturalists to hypothesize that the manganese in the nodules and coatings came not from seawater but from some other source. However, they had differing opinions about what that source might be. John Murray, noticing that nodules were often associated with volcanic rocks or minerals, proposed that the ultimate source was volcanic material. Seafloor alteration of volcanic ash, for example, would release manganese into seawater, and it would quickly precipitate out in oxide form around a shark's tooth or some other object. Murray's hypothesis was strongly influenced by his microscopic examination of sediments brought up in *Challenger*'s dredge hauls; he observed that most grains of volcanic ash in the deep sea mud had been highly altered by contact with seawater. John Buchanan, on the other hand, thought that deep sea animals must be responsible for mobilizing the manganese: as they munched their way through ocean-bottom mud, chemical processes in their guts would release manganese contained in the mineral grains. Buchanan proposed that a complicated series of chemical reactions would then follow before the manganese finally precipitated. Alphonse Renard, the geologist who worked with Murray after the expedition to describe and analyze the *Challenger* sediment samples, was the only scientist who maintained that the manganese must come directly from seawater. He and Murray were close colleagues, but on this point they must have agreed to disagree. Murray dismissed Renard's idea, writing that "so far as our researches go, there are no traces of manganese in solution in the great body of ocean water."

The discovery that manganese nodules are a common feature of the deep sea floor stimulated the curiosity of scientists not directly connected with the *Challenger* expedition, and for many years after the voyage papers appeared in scientific journals proposing a variety of ideas about the source of the manganese and how the nodules formed. One hypothesis held that the manganese originated in undersea springs, another that it came from cosmic dust, a third that the material of the nodules was secreted by bacteria. In 1894, almost twenty years after he had returned from the expedition, Murray and a colleague published a paper defending his original theory, the idea that the manganese came from the alteration of volcanic material. They evaluated and ruled out the other hypotheses one by one, and then concluded: "All subsequent researches on the subject, seem to confirm this view [Murray's theory], which may now be regarded as firmly established." This was not the final word, but as time went on excitement among scientists about the strange objects dissipated. In

A "botryoidal" manganese nodule from a depth of 17,400 feet—more than three miles—in the North Pacific. It is about three inches across, and Murray noted that more than thirty similar nodules were scooped up by the dredge at the station where it was collected. (Murray and Renard, *Report on Deep-Sea Deposits*, plate 2.)

the first half of the twentieth century there were occasional reports of manganese nodules dredged up by oceanographic expeditions, but little attention was paid to them. The nodules were all but forgotten.

Then, after the Second World War they once again emerged as a hot topic of research. It was a time of reinvigorated interest in the oceans generally, and there was an explosion of ocean exploration, especially by American oceanographers who benefited from generous funding by the U.S. Navy. As it became obvious how widespread manganese nodules are on the seafloor, some researchers realized that they might have economic potential. John Buchanan's initial analyses had shown that in addition to being rich in manganese, a metal that is important for making steel, the nodules also contain nickel, copper, and cobalt. It was these other metals that really drew people's attention. Buchanan had reported only trace quantities, but new and more precise analyses revealed that in many nodules nickel and copper were present at levels of 1 to 2 percent, and cobalt at several tenths of a percent. Such small amounts do not sound particularly exciting, but to an economic geologist they are compelling. Ores processed at most mines on land do not have such high concentrations of these elements. Mining the deep sea nodules would pose special logistical problems, but if techniques could be developed to retrieve them easily they would be extremely valuable.

The burst of new research that followed answered many of the questions that had concerned Murray and other nineteenth-century investigators. The *Challenger* scientists' observation that abundant nodules were largely restricted to the deepest parts of the ocean was confirmed as more of the seafloor was sampled and photographed. (Several years after the expedition Buchanan discovered manganese concretions in the relatively shallow water of a Scottish sea loch, throwing doubt on the original observation. But while these are superficially similar to the deep sea nodules, they are not widespread, and their chemical composition is different.) In addition, the rate at which the deep sea nodules grow was put on a quantitative footing using newly developed dating techniques. These showed that the average rate is exceedingly low, in the range of only a few hundredths of an inch, or less, per million years. By comparison, the sediments in the nodule-rich areas are deposited about a thousand times faster. This information about accumulation rates, much more precise than that available to the *Challenger* naturalists, again highlighted the puzzle of why the nodules are not quickly buried in sediment. But scientists today have arrived at an answer to this question that is not radically different from that of the earlier researchers: the nodules are rolled around by currents and bottom animals, and in addition the currents may sweep sediments from their surfaces. Also, biologists have discovered a unique community of small organisms making their home on nodule surfaces. These creatures gobble up much of the falling debris, helping to keep the nodules free of sediment.

Modern analytical methods show that dissolved manganese does occur in seawater. The concentration, as noted earlier, is extremely low, but because of the huge quantity of water in the oceans the total inventory of manganese is enormous. About half this manganese is delivered to the ocean by weathering of continental rocks and transportation by rivers; the other half is supplied by ocean-bottom hot springs, which were not discovered until the 1970s but are now recognized as an important source of many of the elements in solution in seawater. Because both manganese and iron are so readily oxidized, neither element stays dissolved for long. The average atom of manganese probably resides in seawater for less than a thousand years before it precipitates out as manganese oxide; for iron the time scale is even shorter, around a hundred years. Broadly speaking, the input of these elements into the oceans is about the same as the output, so their concentrations in seawater change little over time. Most of the output goes to form manganese nodules or crusts.

Why does much of the manganese end up as nodules instead of falling to the seafloor as a uniform coating? John Murray, observing that nearly every nodule he cut open had a nucleus at its center, concluded that a supply of suitable objects around which nodules could grow was a necessary condition for their occurrence—and such objects are common on many parts of the seafloor. His conclusion anticipated the findings of researchers working more than a century later: surfaces matter. In a general sense surfaces are highly important for chemical reactions, and an object like a pebble of pumice or a whalebone on the ocean bottom serves to increase the precipitation rate of iron and manganese oxides manyfold. Once the process begins, the early formed oxides catalyze even more precipitation. Not only that, they "scavenge," or incorporate other metals, like the nickel, copper, and cobalt Buchanan found in the first few nodules he analyzed. It is also possible that bacteria mediate the deposition of manganese oxides. On a submicroscopic scale, the arrangement of atoms and electrical charges on solid surfaces profoundly affects the types and rates of chemical reactions that occur, and because the manganese and iron oxide grains comprising a nodule are extremely small, they offer a huge surface area for chemical reactions to proceed. In the 1970s one researcher prepared a synthetic batch of manganese oxide of the type that occurs in nodules. He found that its effective surface area was on the order of *several thousand square feet* for every gram (an ounce is 28 grams) of oxide. If this comes close to describing the situation within the nodules, their ability to absorb an array of trace metals from seawater should come as no surprise.

The reinvigorated research on manganese nodules did not focus entirely on questions of basic science; some investigators were equally interested in the nodules' economic potential. Initially this interest centered on the trace metals, particularly cobalt. In 1977, one researcher, an early and consistent proponent of the economic value of manganese nodules, predicted, "It appears that within the next five to ten years, assuming no political and/or legal interferences, the nodules should be in full-scale, economic production as a valuable source of important industrial metals." But fast-forward forty-plus years, and no mining has yet begun. One reason is that the market prices of copper, nickel, and cobalt fell drastically in the late 1970s and the 1980s.

Even in the face of this economic uncertainty, however, mining companies and governments continued to investigate the possibility of mining the nodules. It is estimated that between 1974 and 1982 alone at least one billion dollars was spent exploring and evaluating the potential of mining deep sea manganese nodules. Since then more money has been invested. In the twenty-first century potential miners have also shown an interest in trace elements beyond the nickel-copper-cobalt trio. Those metals are still important, but the nodules, it transpires, are also potential sources of the so-called rare earth elements. These are crucial components of many technologically advanced devices, from mobile phones and computers to catalytic converters and rechargeable batteries. Worldwide demand has skyrocketed since the mid-1990s. And a single country, China, holds most of the world's known supply. Another source would be very welcome.

The most valuable seabed nodule fields are in international waters; for this reason interest in their exploitation has had global consequences. The prospect of countries and private companies competing to mine manganese nodules was one of the driving forces for the United Nations Convention on the Law of the Sea, which met over a number of years until the law was passed and signed in 1982, though it did not come fully into force until 1994, when the requisite sixty nations had ratified it. An intergovernmental organization, the International Seabed Authority (ISA), was then set up to regulate exploration and mining in international waters. Since 2001 the ISA has issued seventeen contracts to national and industrial groups to explore for nodules. Countries as different in size and seagoing expertise as China, Belgium, Russia, and Tonga are represented among the contract holders. Under the permits some groups have already built and tested prototype nodule-recovery devices. With one exception, the contracts are for exploration within a vast but carefully delineated area of the central Pacific that stretches roughly south and east from Hawaii toward Central America. *Challenger*'s most prolific dredge hauls of manganese nodules came from just outside the formal boundaries of this region.

The continued interest in deep-ocean nodule mining has sparked concern about possible environmental consequences, and a number of groups are currently conducting research aimed at understanding how mining might affect the seafloor, particularly its flora and fauna. Although no consensus exists about the type of mining equipment that will be used, most scenarios envision a remotely controlled vehicle that would travel back and forth on the seafloor collecting nodules; the nodules would then be transported to the surface for processing via a hydraulic system. Supporters of deep sea mining argue that this kind of operation would do less damage to the environment than land-based mining, which can entail the clearing of forests, air and water pollution, and sometimes appropriation of land and forced relocation of residents. Opponents point out that mining would have a devastating effect on deep sea creatures that live exclusively in and on the nodules (a fauna that is not yet well characterized) and that because nodules grow so slowly these communities could not easily regenerate once most of the nodules had been removed from a mining site. Organisms living in the top few inches of sediment would also be severely affected by both the passage of collection vehicles and the clouds of disturbed sediment they would create. Even far beyond the mining area, resettling sediment would blanket the seafloor, with potentially serious consequences for ocean-bottom communities. Because flora and fauna are sparse in the deep sea, and biological processes so sluggish, recovery times from any disturbance are likely to be very long.

There is currently little hard information that can help scientists assess the effects of mining. A few test sites where short-term trials (lasting hours or days) of experimental mining equipment have been carried out have been revisited and sampled. Small areas of seafloor have been deliberately disturbed in experiments designed to simulate mining. As might be expected, in most of these cases the effects of the disturbances were negative. Analysis and monitoring showed that both species diversity and the density of organisms decreased. Perhaps the most encouraging aspect of this work is that several groups actively studying the effects of seafloor mining have made recommendations to the ISA that, if followed carefully, will ensure that valuable baseline information is collected during the large-scale test mining that will necessarily precede any full-scale commercial ventures. Ultimately, whether or not manganese nodule mining becomes a reality will depend on economics and the demand for the various elements the nodules contain. But if exploitation does begin, one of the recommendations that has been made to the ISA seems unassailable: tracts of nodule-containing seafloor should be set aside as refuges, off-limits to mining. Such a move would, at a minimum, protect the unique but poorly known ecosystems that depend on the nodules for their survival.

It is never possible to foretell the future, and the *Challenger* scientists could not have imagined how important the manganese nodules they discovered would become. For them, the nodules were scientific curiosities, although John Buchanan did write a letter to his father from *Challenger* in which he described the nodules and suggested that they might be a useful source of manganese at some future time. John Murray is the person most closely associated with collection, curation, and analysis of these peculiar objects, and he maintained an interest in them long after the conclusion of the expedition. That is reflected in an influential book titled *Marine Manganese Deposits* that was published in 1977. Its thirteen chapters, each one written by an eminent scientific expert, describe and analyze different aspects of manganese in the oceans. The frontispiece is a full-page color portrait of Murray. The caption offers no details; it does not even identify the painter. It reads simply, "Sir John Murray, 1841–1914." The assumption is that anyone who reads the book will know who Murray was and understand his key role in the discovery of seafloor manganese accumulations.

The *Challenger* expedition was life-changing for its participants. Henry Moseley, for one, said that when the ship arrived back in England he was "almost sorry" to leave her and "return to the more complicated relations of shore-going life." But the voyage had far-reaching consequences beyond the personal. Both before and for several decades after the expedition other seagoing voyages in small or large part made ocean science part of their mission, and many of these made substantial contributions to what we now know as the science of oceanography. In some ways the *Challenger* expedition simply built on and extended the work of its predecessors and prepared the way for those that followed. But most of those other voyages are not very well known, including to oceanographers. Almost all oceanographers, though, whether their field is geology, chemistry, biology, or physical oceanography, are likely to know something about the *Challenger* expedition. Her voyage has been more frequently written about than others and is more often referred to in scientific papers and articles even today, and she is the ship after which numerous vehicles of exploration and discovery, including a space shuttle, an *Apollo* lunar module, and a deep-ocean-drilling ship, were named. The expedition drew widespread attention even as it was under way, and at its conclusion "scientific men" from around the world wrote to congratulate Wyville Thomson and the entire crew for their accomplishments. Several of these letters appeared in *Nature*. They were full of glowing words for a voyage that had "render[ed] such prominent services to science," and "revealed a New World to Biological Science and opened a new and important field for physical research."

Why does the *Challenger* expedition hold this exalted place in the annals of natural history? There are a number of reasons. Foremost among them is the success of the voyage in meeting its ambitious scientific goals and providing an overall characterization of the deep ocean, from its biology to its physical properties and the nature of the deep seafloor. But the personalities and capabilities of the people involved, and the care and effort put into fleshing out scientific objectives prior to the expedition, were also crucial. Before, during, and after the expedition, reports, letters, and scientific articles about *Challenger*'s exploits and discoveries kept the voyage in the public eye. Thomson, in particular, recognized the importance of communication; during the expedition he sent off reports meant for general consumption as frequently as he could. Were he alive today he would undoubtedly be a prolific blogger. Other *Challenger* participants also wrote about their experiences in books meant for general readers; such publications became a minor cottage industry in the years after the expedition. For decades after the expedition the public interest surrounding *Challenger*'s accomplishments was kept alive by these memoirs. Perhaps most important for *Challenger*'s prominence, though, was the post-expedition attention given to organizing research on the collections and making sure that the results were published together as the Challenger Report. Finally, and not inconsequentially, the people involved in the expedition were in the right place at the right time.

The work of all six scientists whose exploits on board *Challenger* have been detailed in this book were certainly crucial for the success of the expedition. But without three key people—Wyville Thomson, William Carpenter, and George Richards (the hydrographer of the navy)—the voyage might never have happened. And without a fourth, John Murray, *Challenger*'s legacy would not be what it is today. The story began in 1868, when William Carpenter visited Wyville Thomson to discuss their common interest in Crinoids, a group of marine organisms that includes sea lilies and feather stars (named for their multiple feathery arms). Thomson had begun studying these creatures as fossils, but by the time of Carpenter's visit he was also investigating living examples and comparing them with their fossil relatives. Both men were keen to push their work farther through field studies investigating the fauna—Crinoids and otherwise—in the seas around Ireland and Britain, especially the deeper areas away from the immediate coastline.

Such studies would require resources far beyond any they could command themselves. So they hatched a scheme. If Thomson were to write a formal letter outlining their plans for biological work at sea, Carpenter, a gifted speaker and a workaholic, prominent in scientific circles in London, and vice president of the Royal Society, would present it to the Royal Society, recommending that it be submitted to the Admiralty with a request for the loan of a naval ship. Richards had a keen interest in science and had been elected to the Royal Society two years earlier. He and Carpenter were well acquainted. The Thomson-Carpenter proposal appealed to him, and he agreed to give the plan his strong support with the Admiralty. Within a few months a ship had been dispatched, and Thomson and Carpenter's scheme became a reality, an example of the old-boy network at its best.

The 1868 summer cruise and others that followed over the next two seasons were successful, whetting Carpenter's and Thomson's appetites for a global expedition. Again they worked with the Royal Society and Richards to promote the idea. Carpenter in particular was relentless, using his contacts in the government to advance the plan. He and Thomson made a formidable pair; without their groundwork and the support of Richards, the *Challenger* expedition might never have been launched.

Murray's contributions during the expedition and especially after it ensured that *Challenger*'s place in history would be a prominent one. His early experience curating his grandfather's museum made him the ideal person to manage the *Challenger* collections. Murray made sure that the expedition's samples and specimens were described, labeled, and properly stored as soon as they were collected, and when the ship made a port call, he would have them packed in crates and dispatched back to Edinburgh. Later, when he became administrator of the Challenger Office, his organizational and administrative skills guaranteed that nearly every specimen was examined by experts and their descriptions and observations published in the Challenger Report. This may seem straightforward and routine. But Murray's work in the Challenger Office put the expedition on the map in a way that cannot be overestimated. It has often been said that for many years after *Challenger*'s voyage the main contribution of subsequent expeditions was to fill in gaps left by her research, not break new ground.

Advance scientific planning also helped the *Challenger* expedition stand out, something I commented on briefly in the first chapter of this book. Despite the complexity of the voyage, the organizing was completed quite rapidly. Early in 1872, when the Admiralty accepted the Royal Society's proposal and confirmed that a ship would be ready for the expedition to depart by the end of the year, the Royal Society's Circumnavigation Committee immediately swung into action and began planning the scientific aspects of the voyage. They had hundreds of decisions to make. Many were mundane: what type of dredges should be ordered, how many feet of dredging rope would be needed, how many microscopes and specimen bottles would be required, what chemical supplies were necessary. The layout of the on-board chemical and natural history laboratories had to be decided. Different aspects were assigned to committee members with specific experience in those areas. Most exciting would be devising the overarching research plan, which included determining the primary scientific goals and priorities and laying out a tentative route for the ship. Crucially, the committee included in its report that "a full and complete publication of the results of the voyage, with adequate illustrations should form a part of the general plan."

During a nearly three-and-a-half-year voyage there undoubtedly must have been moments of friction, but the overall impression given by the existing reports, journals, and diaries is that the naval officers and the civilian scientists collaborated closely to carry out the scientific plan, and that each party understood well the goals and constraints of the other. *Challenger*'s captains, George Nares and, later, Frank Thomson, made it clear to the crew that they were there to serve the scientific goals. Wyville Thomson later wrote that the sometimes messy scientific tasks—among other things they included dumping dredgeloads of mud or crumbly black manganese nodules onto a recently swabbed deck—were tolerated with "wonderful temper" by the crew, who were "ministers of cleanliness and order."

Communication was vital in giving the expedition wide visibility, a fact recognized by Wyville Thomson. He wrote well, was keen to broadcast the excitement of his science, and was anxious to highlight the value of public funding for large scale scientific projects. He once said that his book *Depths of the Sea* had been written to stimulate public interest in the field of marine science, and also to illustrate that public financing of the cruises it chronicled was fully justified by the advancements made to human knowledge. Throughout *Challenger*'s voyage Thomson found time to send occasional short excerpts from his journal to the popular British monthly magazine *Good Words;* in all, fifteen of these brief essays appeared in print. They were eagerly read by the magazine's followers. After *Challenger* returned, Thomson published a more extensive account of the work done by the expedition in the Atlantic Ocean in a two-volume set called *Voyage of the "Challenger": The Atlantic.* Although quite technical in places, these volumes were intended for general readers. He had planned to publish similar works for the Pacific and Southern Oceans, but illness and his early death in 1882 forestalled them.

Thomson's *Atlantic* appeared in 1877, and it was joined that same year by two additional books about the expedition, both by naval officers and both less technical than Thomson's: *The Cruise of Her Majesty's Ship "Challenger,"* by W. J. J. Spry, and *Log-Letters from "the Challenger,"* by George Campbell. The stream of books about the voyage continued in 1878 when the expedition's artist, J. J. Wild, published *At Anchor,* which concentrated on his experiences ashore and was illustrated with his drawings. A year later, in 1879, Henry Moseley's *Notes by a Naturalist on the "Challenger"* appeared. That so many books about the voyage were published over the space of a few years—and that several of them appeared in multiple editions— suggests that there was a ready market for such accounts. Finally, after a gap of almost sixty years, the last of the books written by a *Challenger* participant appeared: a beautifully illustrated two-volume set of naval officer Herbert Swire's journal, titled *The Voyage of the Challenger*. It was privately published by his son shortly after Swire's death. Together these diverse books reached a wide audience, and, coupled with the numerous news articles that had appeared during the expedition in publications such as the *Times* of London and *Punch*, they kept *Challenger*'s accomplishments in the public eye from the inception of the voyage until long after it had concluded. A clue to the widespread public impact of the *Challenger* expedition can be gleaned from an unlikely source: the writings of Sir Arthur Conan Doyle. In the early 1900s, more than thirty years after the voyage ended, Doyle created an inquisitive, imaginative, world-traveling character who appeared in several of his novels and short stories, reputedly modeled on a combination of an explorer friend of the author's and one of his former professors. Doyle named his character Professor Challenger. His readers would have had no difficulty making the connection with the famous ship and her celebrated expedition.

A coincidence of timing and a confluence of people and events conspired to make the *Challenger* expedition a turning point in ocean research. In the first half of the nineteenth century, leading up to the time of the voyage, most scientific research, especially in Britain, was done by individuals. Some worked in universities with a group of students or assistants. A few, like Charles Darwin, were financially self-sufficient and could support their own work. These scientists belonged to various societies—the Royal Society, the Geological Society, the British Association of Science—at whose meetings they periodically gathered to discuss their research and exchange ideas. But the practice of science was in transition. The model of individuals laboring away in semi-isolation was inadequate for a project like the *Challenger* expedition; such a large-scale enterprise would require public support.

Fortunately, the time was right. Britain was wealthy, it had the world's largest, most experienced navy, its private companies were laying undersea telegraph cables, creating a practical demand for knowledge about the seafloor, and there was public demand for knowledge about the natural world: curiosity, exploration, and learning were seen by the middle class as vehicles for self-improvement. Not least, Darwin's theory of evolution by natural selection was being hotly debated among scientists, in churches, and by the general public. Life in the deep sea and on isolated oceanic islands might provide new and crucial evidence.

These various factors converged to ensure that the *Challenger* project was supported by the government as a standalone enterprise. The expedition and the later work it spawned, especially in the Challenger Office in Edinburgh, fostered a new, more collaborative way of doing science. The participants were loosely organized and independent, but supported by public funds. This was a mode of working that enabled them to accomplish more than had been possible previously, and it ushered in an era in which similar organizational structures sprang up elsewhere.

It might seem a big leap from the *Challenger* expedition to the space age, but as I mentioned in the preface the expedition could be viewed as the *Apollo* project of its day. (There are many differences, of course, and I meant the comparison only in the most general terms.) Both were large-scale, government-funded enterprises that sought to serve multiple ends: the advancement of national pride and security, the pursuit of potential commercial advantages, the search for scientific knowledge. Both were hailed as projects of exploration and discovery, and both caught the public imagination. The science writer William E. Burrows called his book about humankind's quest to explore space *This New Ocean*. Deep space before the space age, like the deep ocean at the time of *Challenger*, was unexplored territory.

The *Challenger* and *Apollo* projects were both characterized by careful pre-mission science planning, a key to their success. *Challenger* naval personnel were chosen for their engineering and scientific knowledge; the *Apollo* astronauts received science training for their missions, and one, Harrison Schmitt, was a Ph.D. geologist. Intermingling of naval and civilian personnel in all aspects of the *Chal-*

A field party returning to the ship after work ashore. (Drawing by Elizabeth Gulland, courtesy of the Centre for Research Collections, University of Edinburgh.)

lenger expedition, unique at the time, was a brilliant strategy that impressed the scientific goals of the voyage on the crew and fostered a sense of common purpose. And for both *Challenger* and *Apollo*, sample analysis and publication of results were recognized as integral parts of the project. In both cases, too, the most highly qualified scientists were sought for the investigations, regardless of their nationality.

With the perspective now of almost a century and a half, what can we say about the impact of the *Challenger* expedition? Many of the tangible results of the voyage have been described in this book: the discovery of cosmic material and manganese nodules on the seafloor;

the mapping out of sediment types throughout the world's oceans; the debunking of the notion that the seafloor is blanketed with primitive protoplasm; the delineation of the shallow ridge that bisects the Atlantic Ocean, now known as the Mid-Atlantic Ridge; the discovery of thousands of new marine species; and the confirmation that life exists at the deepest depths it was possible to sample. John Buchanan declared that the science of oceanography was born on February 15, 1873, the day the expedition's first official observing station was occupied. It is an arresting statement. Buchanan knew that naturalists had been studying the oceans and the life within them long before that particular dredge haul; his point was that marine science had not been recognized as a distinct discipline before the expedition, yet by the end of *Challenger*'s voyage, it had taken its rightful place among the sciences. John Murray was probably one of the first scientists to consider himself an oceanographer. Although he had received many honors, his tombstone, a modest granite slab in the Dean Cemetery in Edinburgh, reads simply, "Sir John Murray, Oceanographer."

The less tangible effects of the expedition are more difficult to characterize. An inkling can be gleaned from a collection of papers published in 1972 for the hundredth anniversary of the *Challenger* voyage. They were written for a conference that took place in Edinburgh that year and published as a special issue of the *Proceedings of the Royal Society of Edinburgh*, the journal that a century earlier had published many *Challenger*-related papers. The full conference title was "Second International Congress on the History of Oceanography." A hundred years after the *Challenger* expedition not only was oceanography a distinct and growing discipline, but the *history* of oceanography was a field in its own right. The congress drew historians and scientists from around the world; in some sense, they all saw themselves as part of *Challenger*'s legacy. Some of the papers dealt directly with aspects of the expedition, but many did not. They covered diverse topics: natural radioactivity in ocean sediments (radioactivity had not even been discovered in the 1870s); the recently inaugurated Deep Sea Drilling Program, which utilized the appro-priately named ship *Glomar Challenger;* and the effects of weather on the routing of North Atlantic shipping. The naturalists on *Challenger,* with their wide interests in almost every aspect of science, would, I think, have felt right at home at the 1972 congress.

The Edinburgh meeting highlighted an undeniable part of *Chal-lenger*'s legacy: formation of an international network of scientists interested in the oceans. The process began shortly after the expe-dition ended; it happened naturally, without any formal organ-ization, but it had a significant effect on the development of oceanography as an independent field of study. In the early years the Challenger Office in Edinburgh was the central hub for the expand-ing community of oceanographers. That was where Murray had shipped about six hundred crates stuffed with rocks, sediments, sea-water, starfish, coral, squid, and more. The collection included nearly seven thousand species of marine organisms, some of which had been dredged from depths of nearly nineteen thousand feet. The office also housed expedition data on water temperatures, currents, and seafloor depth from throughout the world's oceans. This abundance of sam-ples and data for a time made Edinburgh the global center of the new field of oceanography. Domestic and international scientists came and went, examining specimens and meeting with Thomson and Mur-ray. With the encouragement of the two men, many of the research-ers began to send samples from their own ocean voyages to the Challenger Office for examination; some sections of the Challenger Report include data from these samples as well as from the expedi-tion's own specimens. Lengthy letters flowed back and forth between the Edinburgh office and the scientists chosen to investigate expedi-tion samples. In all, seventy-six specialists contributed as authors of the various sections of the Challenger Report; many more helped with work on the collected materials. For some the experience was an inspiration to press the case for oceanographic research in their own institutions or countries. The work was also exhausting. Alex-ander Agassiz, an American biologist who wrote the section of the report on sea urchins, complained to Thomson, "I felt when I got through that I never wanted to see another sea urchin and hoped they would gradually become extinct."

By the time the last volume of the report was published, in 1895, and the Challenger Office closed, a small, informal but influential worldwide community of marine scientists had come into being. To a considerable extent it owed its existence to the expedition. A few years later, in 1899, the king of Sweden invited many of these marine scientists from around the world to Stockholm to attend the first International Conference on the Exploration of the Sea. John Murray, who had recently been knighted by Queen Victoria, headed the British delegation. The conference was a signal that oceanography was now firmly established as a discipline in its own right.

Is it too much to imagine that most of today's ocean research and exploration, from large projects like the International Ocean Discovery Program, which cores deep into seafloor rocks and sediments throughout the world's oceans, to the high-tech research ships of many nations that continually ply the oceans in search of new knowledge, and the manned submersibles that can descend the great depths of the Mariana Trench, are part of *Challenger*'s legacy? Perhaps, but the *Challenger* expedition and its research and publication program indisputably served as a model for a new kind of marine research. Since the expedition, and especially since the middle of the twentieth century, the amount of knowledge that has been gained about our planet and its history by oceanographers is breathtaking. The theory of plate tectonics, an idea that upended geological orthodoxy and provided a coherent picture of how our dynamic planet works, was largely proven by evidence from research at sea. Our understanding of how the earth's temperature gyrated from warm to cold and back again during the last few million years of the Pleistocene Ice Age came initially from studies of deep-ocean sediment cores. So did the understanding of *why* those gyrations occurred when they did. Remarkably, geochemists have even worked out ways to determine the temperature of surface seawater in the past by analyzing the shells of foraminifera—the organisms that formed the bulk of the *Globigerina* ooze repeatedly dredged up by *Challenger*. Studies of the long sediment cores that are now routinely recovered by drilling into the seafloor have given us information about how the earth's environment has varied over roughly the past 200 million years, and about how life in the oceans has coped with those variations—information that is essential for understanding how the planet will respond to the rapid changes in the environment under way today as a result of human activity. And those are just a few highlights.

The last two volumes of the Challenger Report summarized all aspects of the scientific work that had been accomplished during and after the expedition. John Murray had written most of the text himself, and he finally shepherded the volumes to publication nineteen years after the voyage ended. In one way or another *Challenger* had occupied most of his adult life, and he probably had a more comprehensive understanding of the project than anyone else. He had also had time to reflect. He was not a man known for hyperbole, but near the end of the final volume he wrote that the work carried out by the *Challenger* scientists and their colleagues marked "the greatest advance in the knowledge of our planet since the celebrated discoveries of the fifteenth and sixteenth century."

Was this hubris? I think not. The small group of inquisitive scientists who conceived of the expedition and spent more than three years sailing the world's oceans to execute its ambitious plan could not have imagined where their journey would lead, but with the aid of hindsight we view them now as the founding fathers of the modern science of oceanography. Their technology may have been crude by today's standards, but their eyes, ears, intellects, and powers of observation and curiosity about the world were unmatched. *Challenger*'s moment in the sun was a turning point in the long history of humankind's fascination with the sea. For the first time, some of the ocean's deep secrets were revealed. The quest to uncover more of its mysteries continues.

BIBLIOGRAPHY AND FURTHER READING

Works by Challenger's *Scientists*

The fifty-volume Challenger Report is in the public domain and is available online through the Internet Archive at archive.org or the Biodiversity Heritage Library at www.biodiversitylibrary.org. PDF versions of each volume are available for download.

Most of the books and reports by the participants in the expedition are also available online at archive.org and biodiversitylibrary.org.

Campbell, George, Lord. *Log-Letters from "the Challenger."* 4th ed., rev. London: Macmillan, 1877.

Moseley, H. N. *Notes by a Naturalist on the "Challenger."* London: Macmillan, 1879.

Spry, W. J. J. *The Cruise of Her Majesty's Ship "Challenger": Voyages over Many Seas, Scenes in Many Lands.* London: Sampson Low, Marston, Searle and Rivington, 1876.

Swire, Herbert, *The Voyage of the Challenger: A Personal Narrative of the Historic Circumnavigation of the Globe in the Years 1872–1876.* 2 vols. London: Golden Cockerel Press, 1938.

Thomson, C. Wyville. *The Depths of the Sea.* London: Macmillan, 1873.

Thomson, Sir C. Wyville. *The Voyage of the "Challenger": The Atlantic; A Preliminary Account of the General Results of the Exploring Voyage of H.M.S. "Challenger" During the Year 1873 and the Early Part of the Year 1876.* 2 vols. London: Macmillan, 1877.

Thomson, Sir C. Wyville, and John Murray, eds. *Report on the Scientific Results of the Voyage of H.M.S Challenger During the Years 1873–76,* 50 vols. London: Her Majesty's Stationery Office, 1880–1895. The Challenger Report.

Wild, John James. *At Anchor: A Narrative of Experiences Afloat and Ashore During the Voyage of H.M.S. "Challenger" from 1872 to 1876.* London: Marcus Ward, 1878.

Wild, John James. *Thalassa: An Essay on the Depth, Temperature, and Currents of the Ocean.* London: Marcus Ward, 1877.

Internet Resources

Birds of paradise: courting and other behavior from the Cornell Lab of Ornithology at www.birdsofparadiseproject.org. Video.

Coral spawning: Many coral spawning videos are available on YouTube. A segment from *Blue Planet*, filmed on the Great Barrier Reef, can be found at www.youtube.com/watch?v=wsaZ8-I7akg. The entire *Blue Planet II* episode on coral reefs is available at www.dailymotion.com /video/x6vmy98. Video.

Grapsus grapsus (the Sally Lightfoot crab): A short educational video showing these crabs in the Galapagos is available on YouTube (www.youtube .com/watch?v=muMtZHcAzio). Video.

Micrometeorites: Jon Larsen's Project Stardust page is available at www .facebook.com/micrometeorites.

Further Reading

Alberti, Samuel J. M. M., "Conversaziones and the Experience of Science in Victorian England." *Journal of Victorian Culture* 8, no. 2 (2003): 208–230.

Barber, Lynn. *The Heyday of Natural History.* New York: Doubleday, 1980.

Bartish, Igor V., et al. "Phylogeny and Colonization History of *Pringlea antiscorbutica* (Brassicaceae), an Emblematic Endemic from the South Indian Ocean Province." *Molecular Phylogenetics and Evolution* 65 (2012): 748–756.

Burstyn, H. L. "Science Pays Off: Sir John Murray and the Christmas Island Phosphate Industry, 1886–1914." *Social Studies of Science* 5 (1975): 5–34.

Cherel, Y., et al. "Diving Behavior of Female Northern Rockhopper Penguins, *Eudyptes chrysocome moseleyi*, During the Brooding Period at Amsterdam Island (Southern Indian Ocean)." *Marine Biology* 134 (1999): 375–385.

Corfield, Richard. *The Silent Landscape.* Washington, D.C.: Joseph Henry Press, 2003.

Costa Campos, Thomas Ferreira da, et al. "Saint Peter and Saint Paul's Archipelago: Tectonic Uplift of Infra-Crustal Rocks in the Atlantic Ocean." In *Geological and Palaeontological Sites of Brazil*, ed. M. Winge et al., 2005. Available at www.sigep.cprm.gov.br/sitio002/sitio002english.pdf.

Darwin, Charles. *The Structure and Distribution of Coral Reefs.* 1st ed., 1842; 2nd. ed., rev., 1874; 3rd ed., 1879. London: Smith, Elder. All three editions available online at darwin-online.org.uk.

Deacon, Margaret. *Scientists and the Sea, 1650–1900: A Study of Marine Science.* London: Academic Press, 1971.

Deacon, Margaret, Tony Rice, and Colin Summerhayes. *Understanding the Oceans: A Century of Ocean Exploration.* London: UCL Press, 2001.

Deacon, M. B. "A Grounding in Science? John Murray of the *Challenger,* His Grandfather John Macfarlane, and the Macfarlane Museum of Natural History at Bridge of Allan." *Scottish Naturalist* 111 (1999): 225–265.

Edwards, Alasdair, and Roger Lubbock. "The Ecology of Saint Paul's Rocks (Equatorial Atlantic)." *Journal of Zoology: Proceedings of the Zoological Society of London* 200 (1983): 51–69.

Freire, A. S., et al. "Biology of *Grapsus grapsus* (Linnaeus, 1758) (Brachyura, Grapsidae) in the Saint Peter and Saint Paul Archipelago, Equatorial Atlantic Ocean." *Helgoland Marine Research* 65 (2011): 263–273.

Glasby, G. P., ed. *Marine Manganese Deposits.* Amsterdam: Elsevier, 1977.

Gollner, Sabine, et al. "Resilience of Benthic Deep-Sea Fauna to Mining Activities." *Marine Environmental Research* 129 (2017): 76–101.

Haddock, Steven H. D., Mark A. Moline, and James F. Case. "Bioluminescence in the Sea." *Annual Review of Marine Science* 2 (2010): 443–493.

Haeckel, Ernst. *Art Forms in Nature* New York: Dover, 1974. Originally published as *Kunstformen der Natur.* Vienna, 1904.

Herdman, William A. *Founders of Oceanography and Their Work: An Introduction to the Science of the Sea.* London: Edward Arnold, 1923. Available online at archive.org.

Hughes, Terry P., et al. "Global Warming Transforms Coral Reef Assemblages." *Nature* 556 (2018): 492–496.

Huxley, Thomas H. "On a Fossil Bird and a Fossil Cetacean from New Zealand." *Quarterly Journal of the Geological Society of London* 15 (1859) 670–677.

Huxley, Thomas H. "On Some Organisms Living at Great Depths in the North Atlantic Ocean." *Quarterly Journal of Microscopical Science*, n.s. 8 (1868): 203–212.

Linklater, Eric. *The Voyage of the "Challenger."* Garden City, N.Y.: Doubleday, 1972.

Livio, Mario. *Why? What Makes Us Curious.* New York: Simon and Schuster, 2017.

Marin, James G., et al. "Light for All Reasons: Versatility in the Behavioral Repertoire of the Flashlight Fish." *Science* 190 (2017): 74–76.

Mayr, Gerald, Vanesa L. De Pietri, and R. Paul Scofield. "A New Fossil from the Mid-Paleocene of New Zealand Reveals an Unexpected Diversity of World's Oldest Penguins. *Science of Nature* 104 (2017): 9.

Moore, Jerry D. *Visions of Culture: An Introduction to Anthropological Theories and Theorists.* 4th ed. Lanham, Md.: AltaMira Press, 2012.

Murray, James W. "The Surface Chemistry of Hydrous Manganese Dioxide." *Journal of Colloid and Interface Science* 46 (1974): 357–371.

Prum, Richard O. "A Comprehensive Phylogeny of Birds (Aves) Using Targeted Next-Generation DNA Sequencing." *Nature* 526 (2015): 569–573.

Rehbock, Philip F., ed. *At Sea with the Scientifics: The "Challenger" Letters of Joseph Matkin.* Honolulu: University of Hawai'i Press, 1992.

Roberts, Stephen J., et al. "Past Penguin Colony Response to Explosive Volcanism on the Antarctic Peninsula." *Nature Communications,* 2017. doi: 10.1038/ncomms14914.

Rona, Peter A. "The Changing Vision of Marine Minerals." *Ore Geology Reviews* 33 (2008): 618–666.

Rutherford, W. H., ed. "Second International Congress on the History of Oceanography: Challenger Expedition Centenary, Edinburgh, September 12 to 20, 1972. Proceedings 1, 2." *Proceedings of the Royal Society of Edinburgh, Section B, Biological Sciences,* vols. 72–73. Edinburgh: Royal Society of Edinburgh, 1972.

Schlee, Susan. *The Edge of an Unfamiliar World: A History of Oceanography.* New York: Dutton, 1973.

Voosen, Paul. "2.7 Million-Year-Old Ice Opens Window on Past." *Science* 357 (2017): 630–631.

Wagstaff, Steven J., and Françoise Hennion. "Evolution and Biogeography of *Lyallia* and *Hectorella* (Portulacaceae), Geographically Isolated Sisters from the Southern Hemisphere." *Antarctic Science* 19, no. 4 (2007): 417–426.

Whittaker, Robert J., and José María Fernández-Palacios. *Island Biogeography: Ecology, Evolution, and Conservation.* 2nd ed Oxford: Oxford University Press, 2007.

Woodson-Boulton, Amy. "Victorian Museums and Victorian Society." *History Compass* 6, no. 1 (2008): 109–146.

INDEX

Page numbers in italics indicate illustrations.

deep sea organisms, 211,
216–217; fascination with
oceanic islands, 182–183;
on Hawaiian gods, 150; at
Inaccessible Island, 96–98; on
indigenous people of Australia,
133–135, 144–145; on indigenous people of Fiji, 136–143;
on indigenous people of Tonga,
135–136; interest in anthropology, 32–33, 132; on Japan, 151,
153; on kava drinking, 137–139;
on Kerguelen cabbage, 186–187;
motivation for coral studies,
169–170; *Notes by a Naturalist
on the "Challenger,"* 182, 234;
organisms named after, 156; at
Oxford, 30–31; on pile dwellings
in the Philippines, 146–147; at
Saint Paul's Rocks, 86–88;
searches for birds of paradise,
196–199; studies Antarctic
icebergs, 116–117; studies corals,
161–162, 168–170, *169;* studies
Grapsus grapsus, 90–92; studies
Milleporids, 162, 165–166, *167;*
studies rockhopper penguins,
96–98; at Tristan da Cunha,
93–95; on Wesleyan Missionary
Society, 143; work on *Peripatus,*
31–32
Moseleya latistella (coral), 164
Murray, John, 26–28, 164; on
atoll formation, 170–171; and
Christmas Island phosphate
deposits, 29, 171–178; coins
term *oceanography,* 29; collects
live Milleporids, 164–165; as

director of Challenger Office,
26, 176; discovers oceanic
habitat of *Globigerina,* 70–72;
discovers seafloor cosmic dust,
10–13; on distribution of
Antarctic diatom ooze,
121–122; early life, 26–27;
forms association to explore
Christmas Island, 175; founds
Christmas Island phosphate
mine, 29; on global distribution of seafloor sediments,
126–127; joins whaling ship, 28;
philanthropy of, 178; search for
Bathybius, 77–78; sets up
marine laboratory near
Edinburgh, 178; studies
Scotland's lochs, 178; visits
Christmas Island, 177
Museums, Victorian era, *204,*
204–207
Mutation, and evolution, 192–193

Nares, George, 62, 233
National Geographic (magazine),
201
Native people. *See* Indigenous
people
Natural History Museum
(London), 206–207
Natural selection, and evolution,
74, 75, 184, 192, 193, 196,
236
Nature (magazine), 208, 212
Nematocysts, 162
Nettle cells. *See* Nematocysts
New Guinea, 147, 194–195,
200–201